Everyday Mathematics®

Student Math Journal 2

The University of Chicago
School Mathematics Project

 Wright Group

The McGraw·Hill Companies

UCSMP Elementary Materials Component

Max Bell, Director

Authors

Max Bell
John Bretzlauf
Amy Dillard
Robert Hartfield
Andy Isaacs
James McBride, Director

Kathleen Pitvorec
Peter Saecker
Robert Balfanz*
William Carroll*
Sheila Sconiers*

Technical Art

Diana Barrie

First Edition only

Photo Credits

Cover: Bill Burningham/Photography
Photo Collage: Herman Adler Design Group

Contributors

Martha Ayala, Virginia J. Bates, Randee Blair, Donna R. Clay, Vanessa Day, Jean Faszholz, James Flanders, Patti Haney, Margaret Phillips Holm, Nancy Kay Hubert, Sybil Johnson, Judith Kiehm, Carla L. LaRochelle, Deborah Arron Leslie, Laura Ann Luczak, Mary O'Boyle, William D. Pattison, Beverly Pilchman, Denise Porter, Judith Ann Robb, Mary Seymoour, Laura A. Sunseri

 This material is based upon work supported by the National Science Foundation under Grant No. ESI-9252984. Any opinions, findings, and conclusions or recommendations expressed in this material are those of the authors and do not necessarily reflect the views of the National Science Foundation.

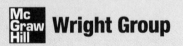

Printed in the United States of America.

Send all inquiries to:
Wright Group/McGraw-Hill
P.O. Box 812960
Chicago, IL 60681

ISBN 0-07-600012-5

19 20 HES 12 11 10

The **McGraw·Hill** Companies

Contents

Unit 7: Fractions and Their Uses; Chance and Probability

Fraction Review	**189**
Math Boxes 7.1	**192**
"Fraction-of" Problems	**193**
Math Boxes 7.2	**195**
Pattern-Block Fractions	**196**
Math Boxes 7.3	**199**
Pattern-Block Fraction Sums and Differences	**200**
Math Boxes 7.4	**201**
Fractions on a Clock Face	**202**
Fraction Addition and Subtraction on a Clock Face	**203**
Math Boxes 7.5	**204**
Picturing Fractions	**205**
Math Boxes 7.6	**206**
Many Names for Fractions	**207**
Hiking	**208**
Math Boxes 7.7	**209**
Fractions and Decimals	**210**
Fraction Name-Collection Boxes	**211**
Math Boxes 7.8	**212**
Comparing Fractions	**213**
Ordering Fractions	**214**
Fraction Problems	**215**
Math Boxes 7.9	**216**
The ONE	**217**
What Is the ONE?	**218**
Math Boxes 7.10	**219**
Making Spinners	**220**
Counting with Fractions	**221**
Math Boxes 7.11	**222**
Expected Spinner Results	**223**
A Cube-Drop Experiment	**224**
Fractions of Sets and Wholes	**226**
Math Boxes 7.12	**227**
Time to Reflect	**228**
Math Boxes 7.13	**229**

Unit 8: Perimeter and Area

Math Boxes 8.1	**230**
Kitchen Layouts and Kitchen Efficiency	**231**
Layout of My Kitchen	**232**
How Efficient Is My Kitchen?	**233**
Work Triangles	**234**
Math Boxes 8.2	**235**
A Floor Plan of My Classroom	**236**
Perimeter	**238**
Areas of Polygons	**239**
Probability	**240**
Math Boxes 8.3	**241**
What Is the Total Area of My Skin?	**242**
Math Boxes 8.4	**244**
Math Boxes 8.5	**245**
Areas of Rectangles	**246**
Scale	**248**
Math Boxes 8.6	**249**
Areas of Parallelograms	**250**
Building a Fence	**253**
Areas of Triangles	**254**
Perimeter and Area	**257**
Math Boxes 8.7	**258**
Math Boxes 8.8	**259**
Comparing Country Areas	**260**
Area	**262**
Time to Reflect	**263**
Math Boxes 8.9	**264**

Unit 9: Percents

Many Names for Percents	**265**
Math Boxes 9.1	**268**
"Percent-of" Number Stories	**269**
Equivalent Fractions, Decimals, and Percents	**270**
Multiplying Whole Numbers	**271**
Math Boxes 9.2	**272**
Math Boxes 9.3	**273**
Discount Number Stories	**274**
Math Boxes 9.4	**275**
Math Boxes 9.5	**276**
Trivia Survey Results	**277**

Dividing Whole Numbers   **278**
Math Boxes 9.6   **279**
Color-Coded Population Maps   **280**
Math Boxes 9.7   **281**
Multiplying Decimals   **282**
Math Boxes 9.8   **284**
Dividing Decimals   **285**
Review: Fractions, Decimals, and Percents   **287**
Math Boxes 9.9   **288**
Time to Reflect   **289**
Math Boxes 9.10   **290**

Unit 10: Reflections and Symmetry

Basic Use of a Transparent Mirror   **291**
Multiplying and Dividing with Decimals   **292**
Math Boxes 10.1   **293**
Presidential Information   **294**
Math Boxes 10.2   **295**
Math Boxes 10.3   **296**
Line Symmetry   **297**
Math Boxes 10.4   **298**
Frieze Patterns   **299**
Multiplying and Dividing with Decimals   **300**
Math Boxes 10.5   **301**
Credits/Debits Game Recording Sheets   **302**
Math Boxes 10.6   **303**
Time to Reflect   **304**
Math Boxes 10.7   **305**

Unit 11: Shapes, Weight, Volume, and Capacity

Estimating Weights in Grams and Kilograms   **306**
Metric and Customary Weight   **307**
Math Boxes 11.1   **308**
Geometric Solids   **309**
Modeling a Rectangular Prism   **310**
Math Boxes 11.2   **311**
Construction of Polyhedrons   **312**
Math Boxes 11.3   **313**
The World's Largest Foods   **314**

Math Boxes 11.4   **315**
Cube-Stacking Problems   **316**
Math Boxes 11.5   **319**
Credits/Debits Game (Advanced Version) Recording Sheets **320**
Math Boxes 11.6   **321**
Converting Measurements   **322**
Math Boxes 11.7   **323**
Time to Reflect   **324**
Math Boxes 11.8   **325**

Unit 12: Rates

Rates   **326**
Math Boxes 12.1   **327**
Rate Tables   **328**
Math Boxes 12.2   **329**
Do These Numbers Make Sense?   **330**
Math Boxes 12.3   **331**
Product Testing   **332**
Unit Prices   **333**
Rates   **334**
Math Boxes 12.4   **335**
Unit Pricing   **336**
More Unit-Pricing Problems   **337**
Math Boxes 12.5   **338**
Looking Back on the World Tour   **339**
Rates   **341**
Math Boxes 12.6   **342**
Time to Reflect   **343**
Math Boxes 12.7   **344**
Route Log   **345**
Route Map   **346**
My Country Notes   **348**
Equivalent Names for Fractions   **356**

Fraction Cards 1   Activity Sheet 5
Fraction Cards 2   Activity Sheet 6
Fraction/Percent Tiles   Activity Sheet 7
Decimal Tiles   Activity Sheet 8

Fraction Review

Divide each shape into equal parts. Color a
fraction of the parts. Write the name of the
"whole" in the **"whole" box.**

1.

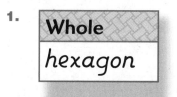

Whole
hexagon

Divide the hexagon into 2 equal parts.
Color $\frac{1}{2}$ of the hexagon.

2.

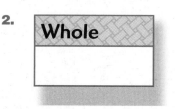

Whole

Divide the hexagon into 3 equal parts.
(*Hint:* Draw 3 rhombuses inside the
hexagon.) Color $\frac{1}{3}$ of the hexagon.

3.

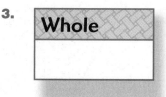

Whole

Divide the rhombus into 2 equal parts.
Color $\frac{0}{2}$ of the rhombus.

4.

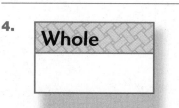

Whole

Divide the trapezoid into 3 equal parts.
Color $\frac{2}{3}$ of the trapezoid.

Fraction Review (cont.)

5.

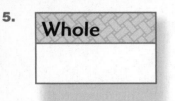

Divide the hexagon into 6 equal parts.
Color $\frac{5}{6}$ of the hexagon.

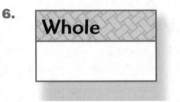

6.

Divide each hexagon into thirds.
Color $1\frac{2}{3}$ hexagons.

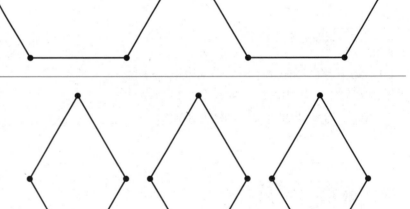

7.

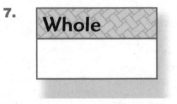

Divide each rhombus into 2 equal parts.
Color $2\frac{1}{2}$ rhombuses.

---

8. Grace was asked to color $\frac{2}{3}$ of a hexagon.
This is what she did. What's wrong?

_____

_____

_____

_____

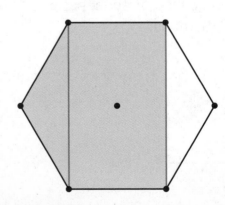

Fraction Review (cont.)

Fill in the missing numbers on the number lines.

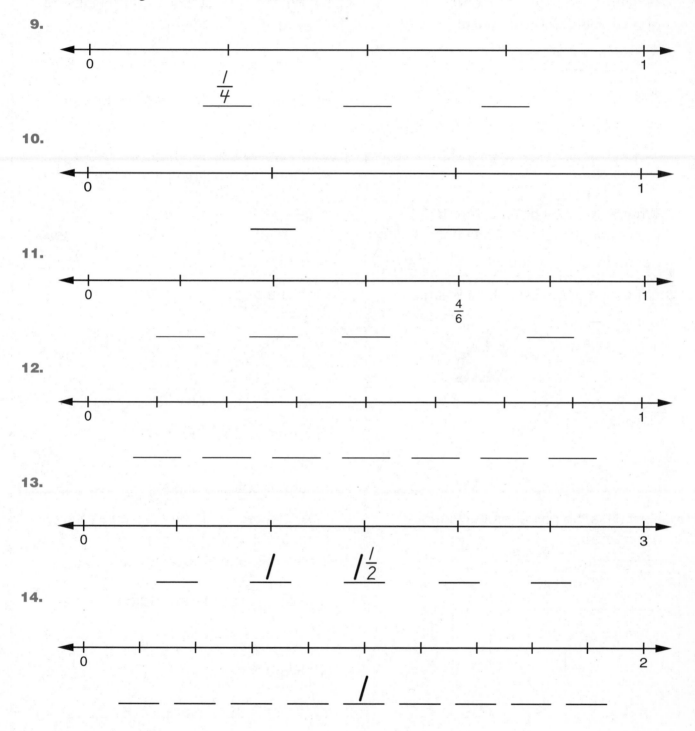

9.

10.

11.

12.

13.

14.

Math Boxes 7.1

1. a. Measure and record the length of each side of the rectangle.

_____ in.

_____ in.   _____ in.

_____ in.

b. What is the total distance around the rectangle called? Circle one.

perimeter   area

SRB 111

2. a. What city is located at 30° N latitude and 10° W longitude?

_____

b. In which country is the city located?

_____

c. On which continent is the city located?

_____

SRB 216 217

3. What fraction of the clock face is shaded?

_____

SRB 54

4. Measure angle POL.

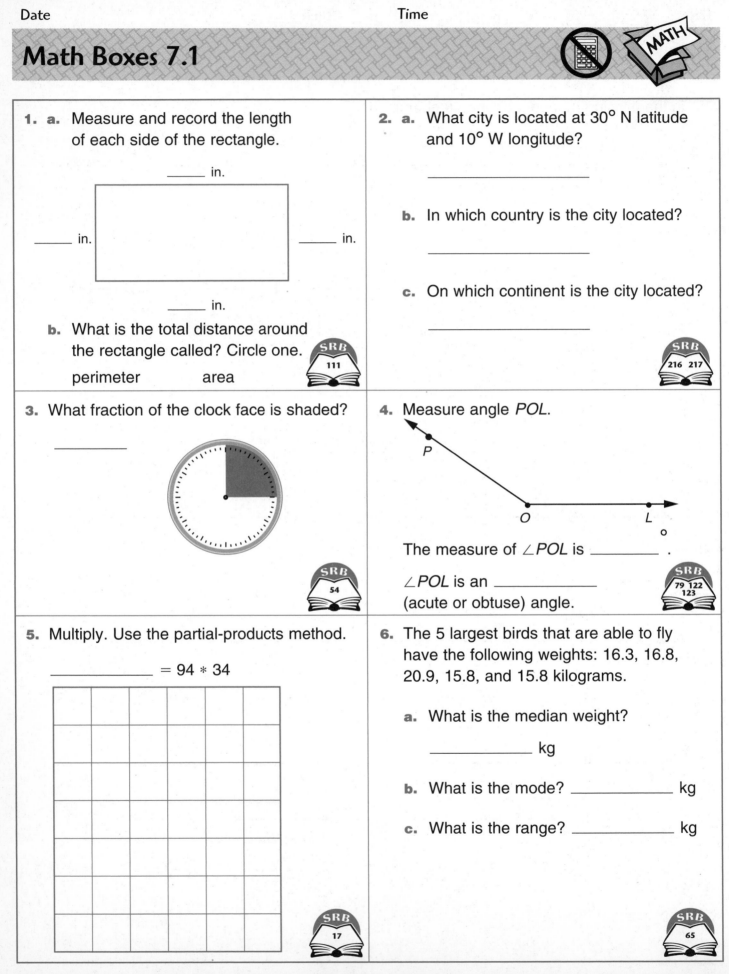

P

O   L

°

The measure of ∠POL is _____ .

∠POL is an _____ (acute or obtuse) angle.

SRB 79 122 123

5. Multiply. Use the partial-products method.

_____ = 94 * 34

SRB 17

6. The 5 largest birds that are able to fly have the following weights: 16.3, 16.8, 20.9, 15.8, and 15.8 kilograms.

a. What is the median weight?

_____ kg

b. What is the mode? _____ kg

c. What is the range? _____ kg

SRB 65

"Fraction-of" Problems

1.

Whole
16 nickels

a. Cross out $\frac{1}{4}$ of the nickels.

b. Circle $\frac{3}{4}$ of the nickels.
   How much money is that?

   $_____._____

2.

Whole

a. Fill in the "whole" box.

b. Circle $\frac{5}{6}$ of the dimes.
   How much money is that?

   $_____._____

3.

Whole

a. Fill in the "whole" box.

b. Circle $\frac{3}{5}$ of the quarters.
   How much money is that?

   $_____._____

"Fraction-of" Problems (cont.)

4. Michael had 20 baseball cards. He gave $\frac{1}{5}$ of them to his friend, Alana, and $\frac{2}{5}$ to his brother, Dean.

a. How many baseball cards did he give to Alana? _____ cards

b. How many did he give to Dean? _____ cards

c. How many did he keep for himself? _____ cards

Solve.

5. $\frac{1}{3}$ of 12 = _____

6. $\frac{2}{3}$ of 12 = _____

7. $\frac{3}{5}$ of 15 = _____

8. $\frac{3}{4}$ of 36 = _____

9. $\frac{5}{8}$ of 32 = _____

10. $\frac{4}{6}$ of 24 = _____

11. $\frac{2}{5}$ of 30 = _____

12. $\frac{5}{6}$ of 30 = _____

13. $\frac{2}{4}$ of 14 = _____

14. What is $\frac{1}{2}$ of 25? _____ Explain.

_____

_____

15. Maurice spent $\frac{1}{2}$ of his money on lunch. He had $2.50 left. How much money did he start with? _____

16. Erika spent $\frac{3}{4}$ of her money on lunch. She had $2.00 left. How much money did she start with? _____

Math Boxes 7.2

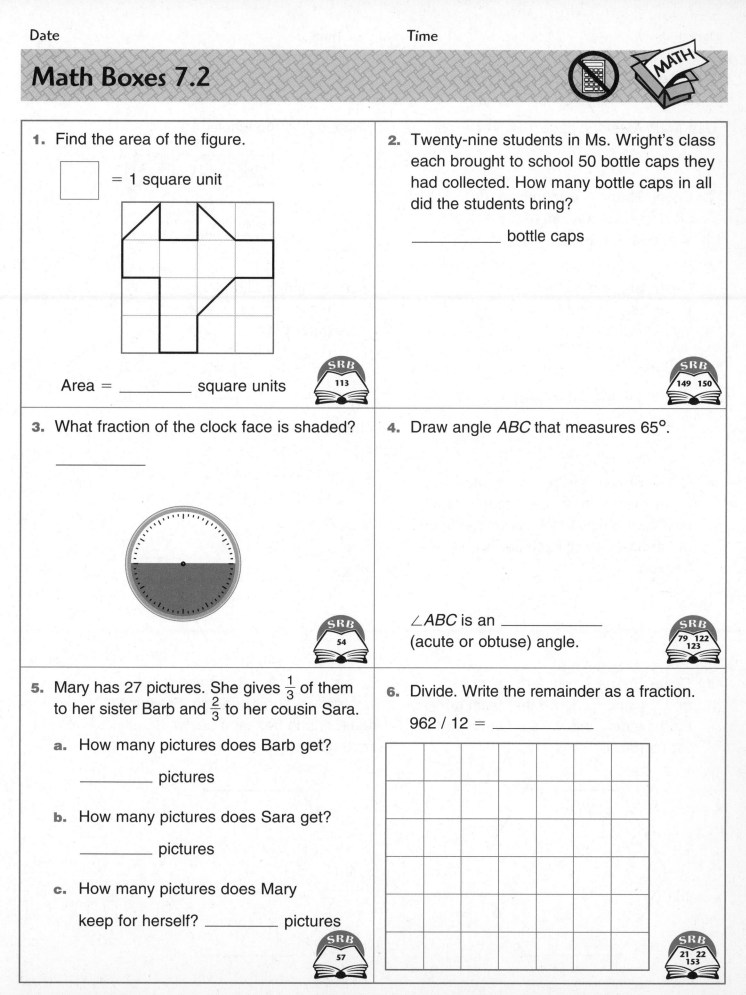

1. Find the area of the figure.

☐ = 1 square unit

Area = _____ square units

SRB 113

2. Twenty-nine students in Ms. Wright's class each brought to school 50 bottle caps they had collected. How many bottle caps in all did the students bring?

_____ bottle caps

SRB 149 150

3. What fraction of the clock face is shaded?

_____

SRB 54

4. Draw angle *ABC* that measures 65°.

∠*ABC* is an _____
(acute or obtuse) angle.

SRB 79 122 123

5. Mary has 27 pictures. She gives $\frac{1}{3}$ of them to her sister Barb and $\frac{2}{3}$ to her cousin Sara.

a. How many pictures does Barb get?

_____ pictures

b. How many pictures does Sara get?

_____ pictures

c. How many pictures does Mary keep for herself? _____ pictures

SRB 57

6. Divide. Write the remainder as a fraction.

962 / 12 = _____

SRB 21 22 153

Pattern-Block Fractions

Use *Math Masters,* page 104. For Problems 1–6, Shape A is the whole.

Whole
Shape A: small hexagon

1. Cover Shape A with trapezoid blocks. What fraction of the shape is covered by 1 trapezoid? _____

2. Cover Shape A with rhombuses. What fraction of the shape is covered by

    1 rhombus? _____

    2 rhombuses? _____

3. Cover Shape A with triangles. What fraction of the shape is covered by

    1 triangle? _____

    3 triangles? _____

    5 triangles? _____

4. Cover Shape A with 1 trapezoid and 3 triangles. With a straightedge, draw how your shapes look on the hexagon at the right. Label each part with a fraction.

5. Cover Shape A with 2 rhombuses and 2 triangles. Draw the result on the hexagon below. Label each part with a fraction.

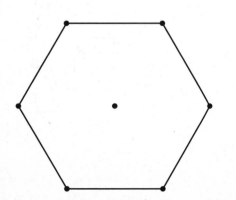

6. Cover Shape A with 1 trapezoid, 1 rhombus, and 1 triangle. Draw the result on the hexagon below. Label each part with a fraction.

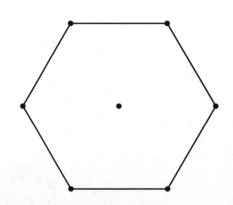

Use with Lesson 7.3.

Pattern-Block Fractions (cont.)

Use *Math Masters,* page 104. For Problems 7–12, Shape B is the whole.

Whole
Shape B: double hexagon

7. Cover Shape B with trapezoids.
   What fraction of the shape is covered by

   1 trapezoid? _____   2 trapezoids? _____   3 trapezoids? _____

8. Cover Shape B with rhombuses. What fraction of the shape is covered by

   1 rhombus? _____   3 rhombuses? _____   5 rhombuses? _____

9. Cover Shape B with triangles. What fraction of the shape is covered by

   1 triangle? _____   2 triangles? _____   3 triangles? _____

10. Cover Shape B with hexagons. What fraction of the shape is covered by

    1 hexagon? _____   2 hexagons? _____

11. Cover Shape B completely
    with 1 hexagon, 1 rhombus,
    1 triangle, and 1 trapezoid.
    Draw the result on the figure
    at the right. Label each part
    with a fraction.

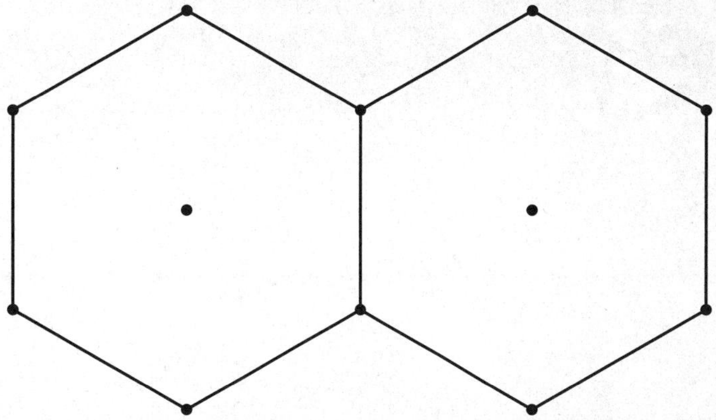

12. Cover Shape B completely
    with 1 trapezoid, 2 rhombuses,
    and 5 triangles. Draw the
    result on the figure at the right.
    Label each part with a fraction.

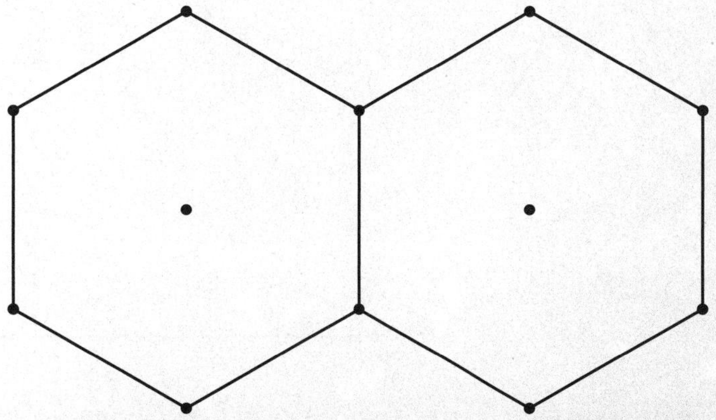

Pattern-Block Fractions (cont.)

Use *Math Masters,* page 104. For Problems 13–16, Shape C is the whole.

Whole
Shape C: big hexagon

13. Cover Shape C with trapezoids.
What fraction of the shape is covered by

    1 trapezoid? _____   2 trapezoids? _____   6 trapezoids? _____

14. Cover Shape C with rhombuses. What fraction of the shape is covered by

    1 rhombus? _____   3 rhombuses? _____   6 rhombuses? _____

15. Cover Shape C with triangles. What fraction of the shape is covered by

    1 triangle? _____   3 triangles? _____   12 triangles? _____

16. Cover Shape C completely, using one or more trapezoids, rhombuses,
triangles, and hexagons. Draw the result on the big hexagon below.
Label each part with a fraction.

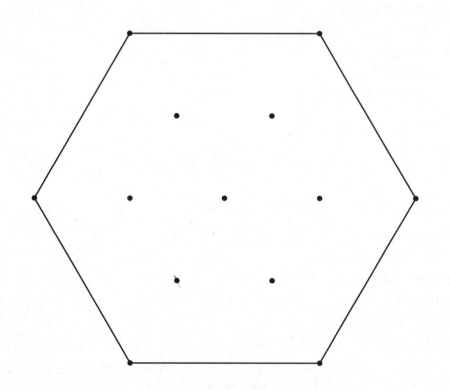

Math Boxes 7.3

1. Measure the length and width of your journal to the nearest half-inch. Find its perimeter.

   a. Length = _____ inches

   b. Width = _____ inches

   c. Perimeter = _____ inches

2. a. What city is located at 60° N latitude and 5° E longitude?

   _____

   b. In which country is the city located?

   _____

   c. On which continent is the city located?

   _____

3. What fraction of the clock face is shaded?

   _____

4. Measure angle MRS.

   The measure of ∠MRS is _____°.

   ∠MRS is an _____ angle.

5. Multiply. Use the partial-products method.

   _____ = 19 * 473

6. Cleo's friends ran the 50-yard dash in the following times:

   7.9, 11, 9.9, 8.3, 10.5, 7.9, and 7.4 seconds.

   a. What is the median time?

   _____ seconds

   b. What is the mode? _____ seconds

   c. What is the range? _____ seconds

Pattern-Block Fraction Sums and Differences

1. Use pattern blocks to find fractions that add up to 1 whole. Draw lines to show the blocks you used. Write a number model to show that the sum of your fractions is 1.

Whole
hexagon

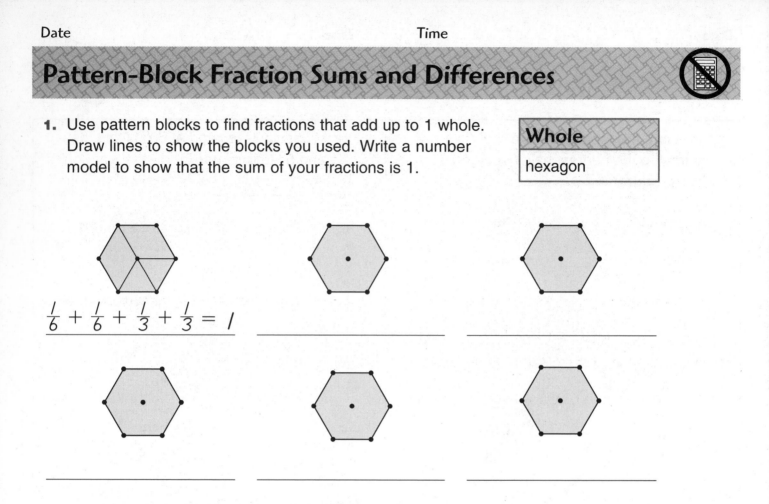

$\frac{1}{6} + \frac{1}{6} + \frac{1}{3} + \frac{1}{3} = 1$

_____   _____

_____   _____   _____

2. Use pattern blocks to find fractions that add up to $\frac{2}{3}$. Draw lines to show the blocks you used. Write a number model to show that the sum of your fractions is $\frac{2}{3}$.

_____   _____   _____

Solve. You may use pattern blocks or any other method.

3. $\frac{2}{3} - \frac{1}{6} =$ _____

4. $\frac{5}{6} - \frac{1}{2} =$ _____

5. $1\frac{1}{6} - \frac{1}{3} =$ _____

6. $1\frac{1}{2} - \frac{5}{6} =$ _____

1. Find the area of the figure.

   ☐ = 1 square unit

   Area = _____ square units

2. According to a recent survey at Star Elementary School, each student eats an average of 17 pieces of candy or servings of junk food per week. About how many pieces of candy or servings of junk food would this be per week for a class of 32 students?

   _____ pieces or servings

3. What fraction of the clock face is shaded?

   _____

4. Draw angle *LMN* that measures 120°.

   ∠*LMN* is an _____ (acute or obtuse) angle.

5. a. In December, $\frac{3}{4}$ of a foot of snow fell on Wintersville. How many inches of snow is this?

   _____ inches

   b. Tina's daughter will be $\frac{5}{6}$ of a year old next week. How many months old will she be?

   _____ months

6. Divide. Write the remainder as a fraction.

   809 / 13 = _____

Fractions on a Clock Face

Write the fraction represented on each clock face.

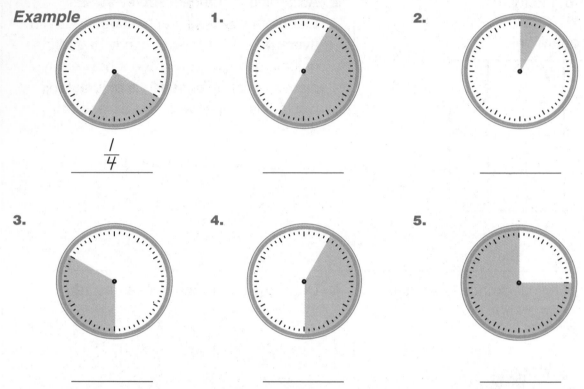

Example

$\dfrac{1}{4}$

1. _____

2. _____

3. _____

4. _____

5. _____

Shade each clock face to show the fraction. Start shading at the line segment.

Example

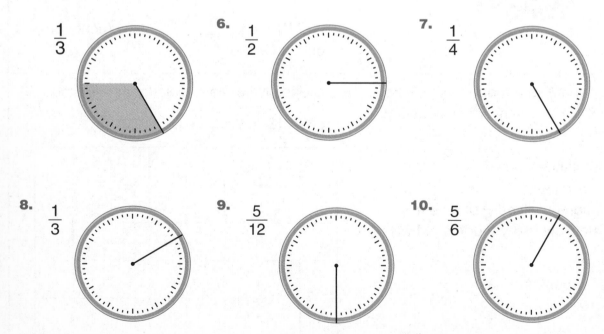

$\dfrac{1}{3}$

6. $\dfrac{1}{2}$

7. $\dfrac{1}{4}$

8. $\dfrac{1}{3}$

9. $\dfrac{5}{12}$

10. $\dfrac{5}{6}$

Use with Lesson 7.5.

Fraction Addition and Subtraction on a Clock Face

Write the fraction addition problem shown on each clock face.

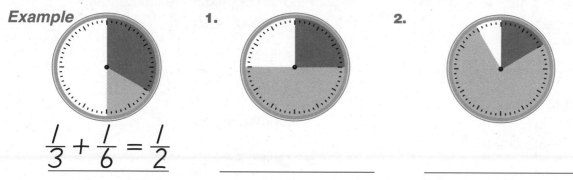

Example   **1.**   **2.**

$$\frac{1}{3} + \frac{1}{6} = \frac{1}{2}$$

_____   _____

Use the clock face to help you solve these addition problems.

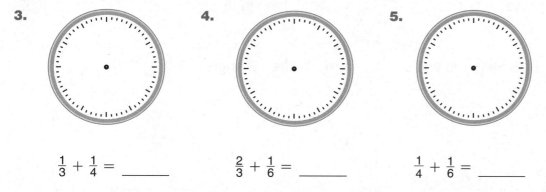

3.   **4.**   **5.**

$\frac{1}{3} + \frac{1}{4} =$ _____ $\frac{2}{3} + \frac{1}{6} =$ _____ $\frac{1}{4} + \frac{1}{6} =$ _____

Write the fraction subtraction problem shown on each clock face.

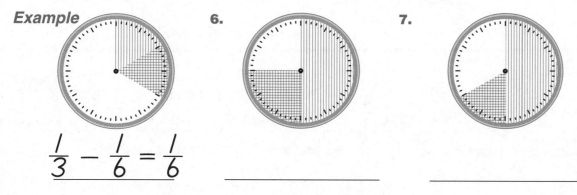

Example   **6.**   **7.**

$$\frac{1}{3} - \frac{1}{6} = \frac{1}{6}$$

_____   _____

Use the clock face to help you solve these subtraction problems.

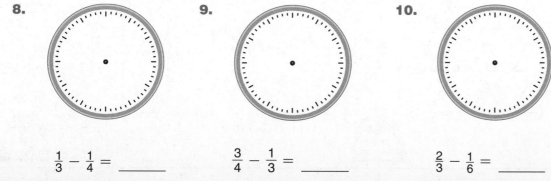

8.   **9.**   **10.**

$\frac{1}{3} - \frac{1}{4} =$ _____ $\frac{3}{4} - \frac{1}{3} =$ _____ $\frac{2}{3} - \frac{1}{6} =$ _____

Math Boxes 7.5

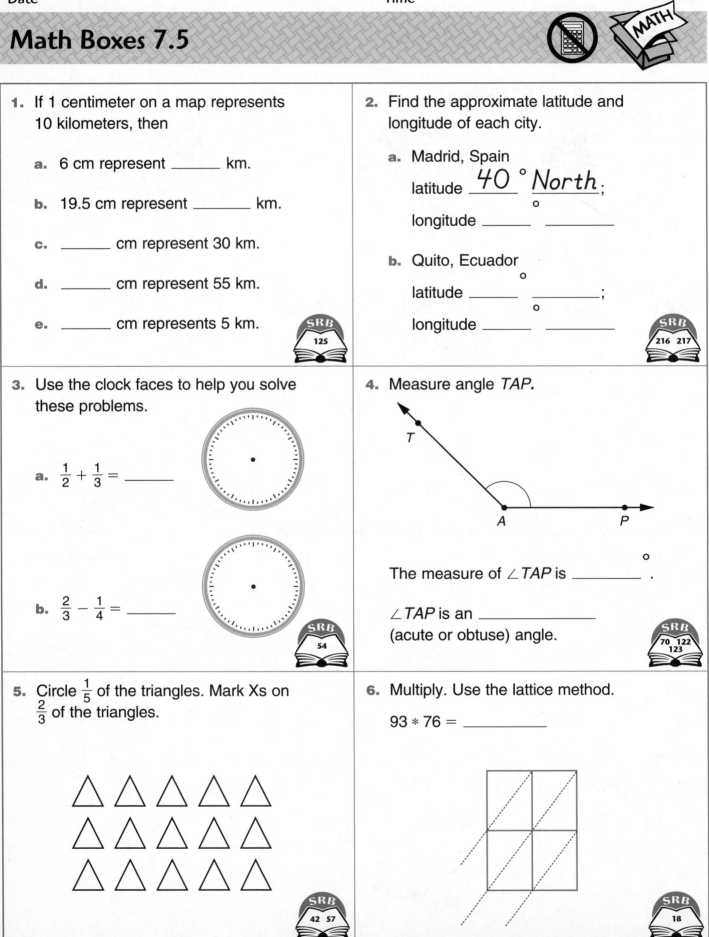

1. If 1 centimeter on a map represents 10 kilometers, then

   a. 6 cm represent _____ km.

   b. 19.5 cm represent _____ km.

   c. _____ cm represent 30 km.

   d. _____ cm represent 55 km.

   e. _____ cm represents 5 km.

   SRB 125

2. Find the approximate latitude and longitude of each city.

   a. Madrid, Spain

   latitude _40_ ° _North_ ;

   longitude _____ ° _____

   b. Quito, Ecuador

   latitude _____ ° _____ ;

   longitude _____ ° _____

   SRB 216 217

3. Use the clock faces to help you solve these problems.

   a. $\frac{1}{2} + \frac{1}{3} =$ _____

   b. $\frac{2}{3} - \frac{1}{4} =$ _____

   SRB 54

4. Measure angle *TAP*.

   T

   A   *P*

   The measure of ∠ *TAP* is _____ °.

   ∠ *TAP* is an _____ (acute or obtuse) angle.

   SRB 70 122 123

5. Circle $\frac{1}{5}$ of the triangles. Mark Xs on $\frac{2}{3}$ of the triangles.

   SRB 42 57

6. Multiply. Use the lattice method.

   93 * 76 = _____

   SRB 18

Use with Lesson 7.5.

Picturing Fractions

1. Circle $\frac{1}{2}$ of the stars. Cross out $\frac{1}{3}$ of the stars that are not circled. Put a box around $\frac{1}{4}$ of the stars that are not circled or crossed out.

   How many stars are left?

   _____ stars

2. Circle $\frac{2}{3}$ of the squares below.

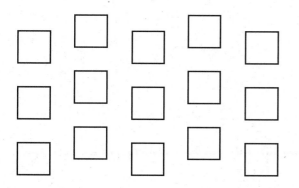

3. Shade $\frac{2}{3}$ of the circle below.

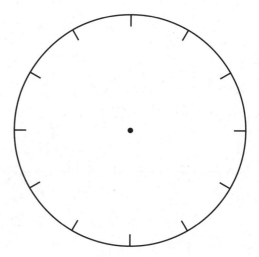

4. How many minutes pass when the minute hand on a clock makes

   a. $\frac{1}{2}$ of a turn? _____ minutes

   b. $\frac{3}{4}$ of a turn? _____ minutes

   c. $\frac{2}{3}$ of a turn? _____ minutes

   d. $\frac{1}{6}$ of a turn? _____ minutes

Math Boxes 7.6

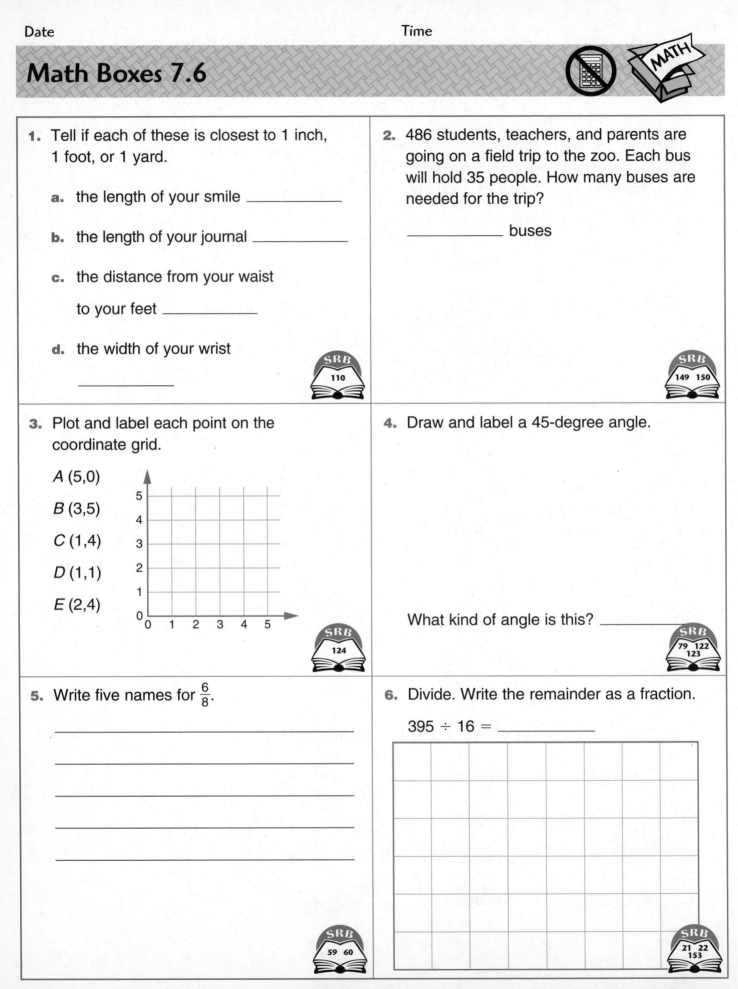

1. Tell if each of these is closest to 1 inch, 1 foot, or 1 yard.

a. the length of your smile _____

b. the length of your journal _____

c. the distance from your waist

to your feet _____

d. the width of your wrist

_____

SRB
110

2. 486 students, teachers, and parents are going on a field trip to the zoo. Each bus will hold 35 people. How many buses are needed for the trip?

_____ buses

SRB
149 150

3. Plot and label each point on the coordinate grid.

A (5,0)

B (3,5)

C (1,4)

D (1,1)

E (2,4)

SRB
124

4. Draw and label a 45-degree angle.

What kind of angle is this? _____

SRB
79 122
123

5. Write five names for $\frac{6}{8}$.

_____

_____

_____

_____

_____

SRB
59 60

6. Divide. Write the remainder as a fraction.

$395 \div 16 =$ _____

SRB
21 22
153

Use with Lesson 7.6.

Many Names for Fractions

Color the squares and write the missing numerators.

1. Color $\frac{1}{2}$ of each large square.

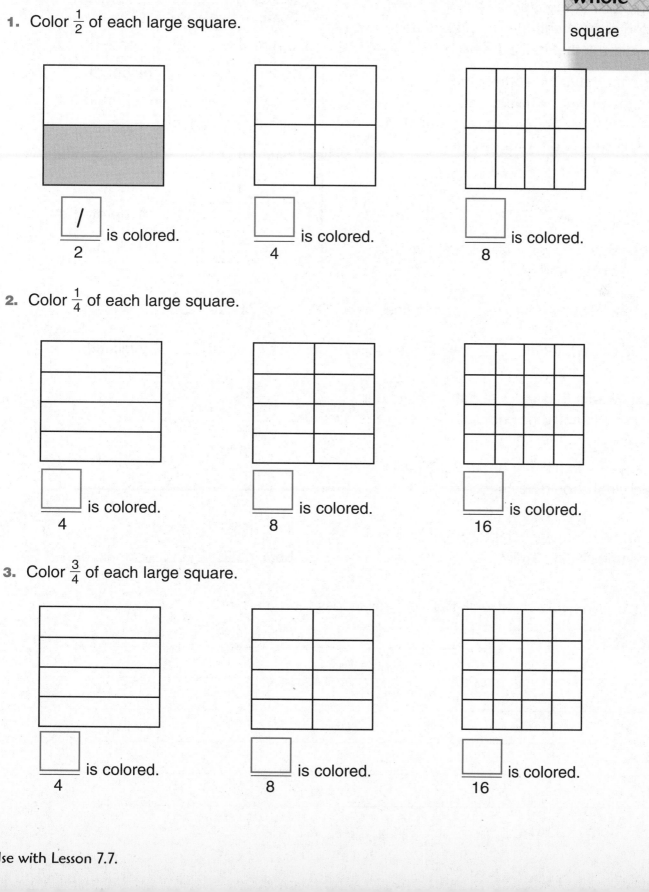

$\dfrac{/}{2}$ is colored. $\dfrac{}{4}$ is colored. $\dfrac{}{8}$ is colored.

2. Color $\frac{1}{4}$ of each large square.

$\dfrac{}{4}$ is colored. $\dfrac{}{8}$ is colored. $\dfrac{}{16}$ is colored.

3. Color $\frac{3}{4}$ of each large square.

$\dfrac{}{4}$ is colored. $\dfrac{}{8}$ is colored. $\dfrac{}{16}$ is colored.

Whole
square

Hiking

Luis is staying in a large state park that has 8 hiking trails. In the table at the right, each trail is labeled easy, moderate, or rugged, depending on how difficult that trail is for hiking.

Luis figures that it would take him about 20 minutes to walk 1 mile on an easy trail, about 30 minutes on a moderate trail, and about 40 minutes on a rugged trail.

State Park Trails

Trail	Miles	Type
Ice Age	$1\frac{1}{4}$	easy
Kettle	2	moderate
Pine	$\frac{3}{4}$	moderate
Bluff	$1\frac{3}{4}$	rugged
Cliff	$\frac{3}{4}$	rugged
Oak	$1\frac{1}{2}$	easy
Sky	$1\frac{1}{2}$	moderate
Badger	$3\frac{1}{2}$	moderate

1. About how long will it take Luis to walk the following trails?

   a. Kettle Trail: About _____ minutes   b. Cliff Trail: About _____ minutes

   c. Oak Trail: About _____ minutes   d. Bluff Trail: About _____ minutes

2. If Luis wants to hike for about $\frac{3}{4}$ of an hour, which trail should he choose?   _____

3. If he wants to hike for about 25 minutes, which trail should he choose?   _____

4. About how long would it take him to complete Pine Trail?   About _____ minutes

5. Do you think Luis could walk Badger Trail in less than 2 hours? _____

   Explain. _____

   _____

   _____

   _____

   _____

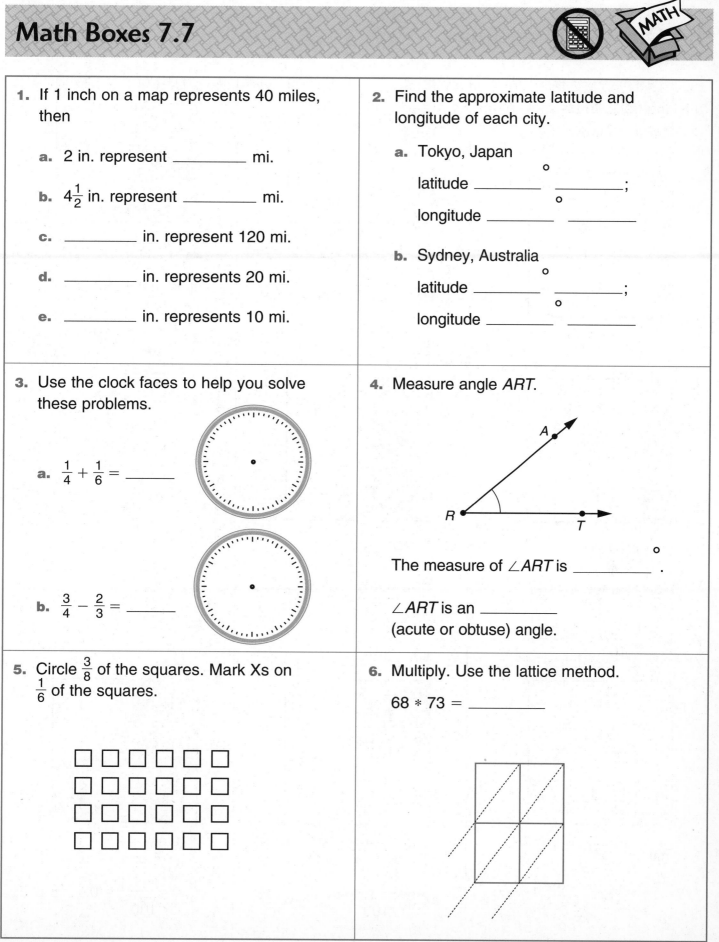

1. If 1 inch on a map represents 40 miles, then

   a. 2 in. represent _____ mi.

   b. $4\frac{1}{2}$ in. represent _____ mi.

   c. _____ in. represent 120 mi.

   d. _____ in. represents 20 mi.

   e. _____ in. represents 10 mi.

2. Find the approximate latitude and longitude of each city.

   a. Tokyo, Japan

    latitude _____ ° _____ ;

    longitude _____ ° _____

   b. Sydney, Australia

    latitude _____ ° _____ ;

    longitude _____ ° _____

3. Use the clock faces to help you solve these problems.

   a. $\frac{1}{4} + \frac{1}{6} = $ _____

   b. $\frac{3}{4} - \frac{2}{3} = $ _____

4. Measure angle *ART*.

The measure of $\angle ART$ is _____ °.

$\angle ART$ is an _____ (acute or obtuse) angle.

5. Circle $\frac{3}{8}$ of the squares. Mark Xs on $\frac{1}{6}$ of the squares.

6. Multiply. Use the lattice method.

68 * 73 = _____

Fractions and Decimals

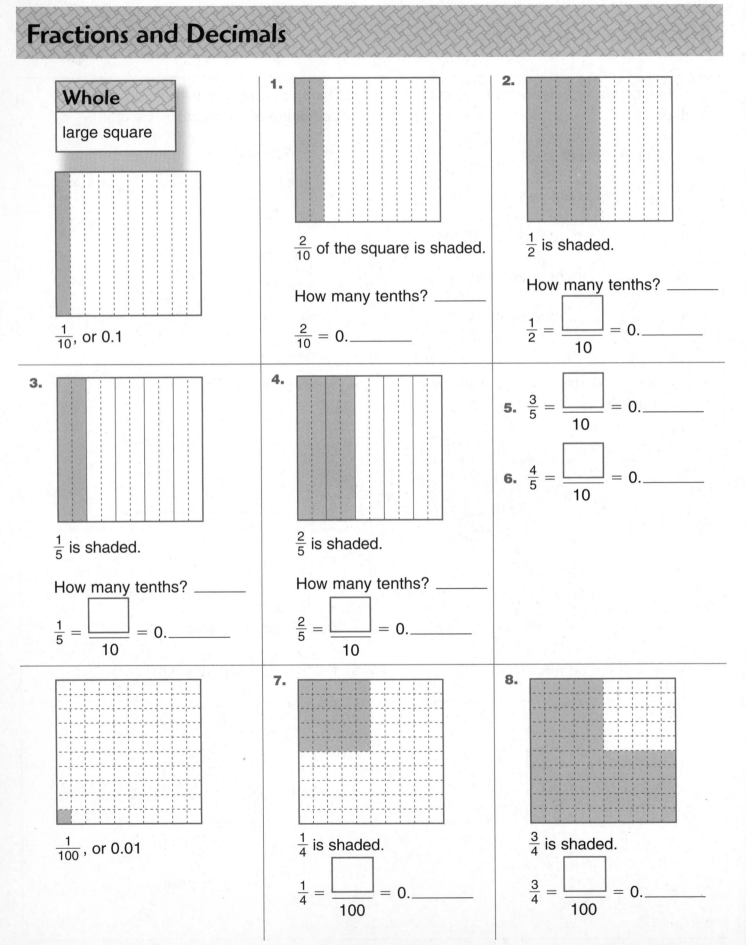

Whole

large square

$\frac{1}{10}$, or 0.1

1.

$\frac{2}{10}$ of the square is shaded.

How many tenths? _____

$\frac{2}{10} = 0.$_____

2.

$\frac{1}{2}$ is shaded.

How many tenths? _____

$\frac{1}{2} = \dfrac{\boxed{}}{10} = 0.$_____

3.

$\frac{1}{5}$ is shaded.

How many tenths? _____

$\frac{1}{5} = \dfrac{\boxed{}}{10} = 0.$_____

4.

$\frac{2}{5}$ is shaded.

How many tenths? _____

$\frac{2}{5} = \dfrac{\boxed{}}{10} = 0.$_____

5. $\frac{3}{5} = \dfrac{\boxed{}}{10} = 0.$_____

6. $\frac{4}{5} = \dfrac{\boxed{}}{10} = 0.$_____

$\frac{1}{100}$, or 0.01

7.

$\frac{1}{4}$ is shaded.

$\frac{1}{4} = \dfrac{\boxed{}}{100} = 0.$_____

8.

$\frac{3}{4}$ is shaded.

$\frac{3}{4} = \dfrac{\boxed{}}{100} = 0.$_____

                                          Use with Lesson 7.8.

Fraction Name-Collection Boxes

In each name-collection box:

- Write the missing number in each fraction so that the fraction belongs in the box.

- Write two more fractions that can go in the box.

1.

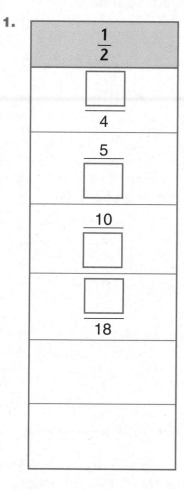

2.

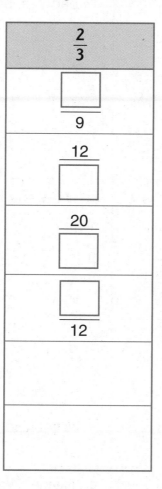

3.

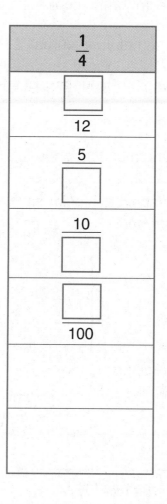

4. Make up your own name-collection box problems like the ones above. Ask a friend to solve your problems. Check your friend's work.

a.

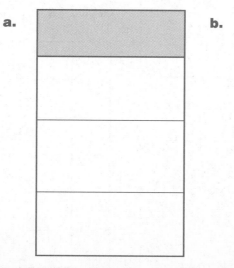

b.

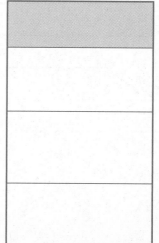

Math Boxes 7.8

1. Tell if each of these is closest to 1 inch, 1 foot, or 1 yard.

   a. the height of the door _____

   b. the width of your journal _____

   c. the length of your largest toe

   _____

   d. the length of your shoe _____

2. There are 252 pages in the book Mike is reading for his book report. He has two weeks to read the book. About how many pages should he read each day?

   _____ pages

3. Plot and label each point on the coordinate grid.

   A (0,2)

   B (4,0)

   C (1,5)

   D (5,5)

   E (5,3)

4. Draw and label a 125-degree angle.

   What kind of angle is this? _____

5. Write five names for $\frac{4}{5}$.

   _____

   _____

   _____

   _____

6. Divide. Write the remainder as a fraction.

   456 / 14 = _____

Use with Lesson 7.8.

Comparing Fractions

Math Message: Eating Fractions

Quinn, Nancy, Diego, Paula, and Kiana were given 4 chocolate bars to share.
All 4 bars were the same size.

1. Quinn and Nancy shared a chocolate bar. Quinn ate $\frac{1}{4}$ of the bar, and Nancy ate $\frac{2}{4}$.

   Who ate more? _____

   How much of the bar was left? _____

2. Diego, Paula, and Kiana each ate part of the other chocolate bars. Diego ate $\frac{2}{3}$ of a bar, Paula ate $\frac{2}{5}$ of a bar, and Kiana ate $\frac{5}{6}$ of a bar.

   Who ate more, Diego or Paula? _____

   How do you know? _____

   _____

Comparing Fractions with $\frac{1}{2}$

Turn your Fraction Cards fraction-side up. Sort them into three piles:

- fractions less than $\frac{1}{2}$
- fractions equal to $\frac{1}{2}$
- fractions greater than $\frac{1}{2}$

You can turn the cards over to check your work. When you are finished,
write the fractions in each pile in the correct box below.

Less than $\frac{1}{2}$	Equal to $\frac{1}{2}$	Greater than $\frac{1}{2}$

Use with Lesson 7.9.

Ordering Fractions

Write the fractions in order from smallest to largest.

1. $\dfrac{1}{4}$, $\dfrac{1}{2}$, $\dfrac{1}{9}$, $\dfrac{1}{5}$, $\dfrac{1}{100}$

_____   _____   _____   _____   _____
smallest   **largest**

2. $\dfrac{2}{4}$, $\dfrac{2}{2}$, $\dfrac{2}{9}$, $\dfrac{2}{5}$, $\dfrac{2}{100}$

_____   _____   _____   _____   _____
smallest   **largest**

3. $\dfrac{4}{10}$, $\dfrac{7}{10}$, $\dfrac{8}{10}$, $\dfrac{2}{10}$, $\dfrac{1}{10}$

_____   _____   _____   _____   _____
smallest   **largest**

4. $\dfrac{4}{25}$, $\dfrac{1}{25}$, $\dfrac{7}{8}$, $\dfrac{6}{12}$, $\dfrac{7}{15}$

_____   _____   _____   _____   _____
smallest   **largest**

5. a. Write 5 fractions that all have the same denominator.

_____   _____   _____   _____   _____

b. Ask a partner to put them in order from smallest to largest.

_____   _____   _____   _____   _____
smallest   **largest**

c. Do you agree with your partner's answer? _____

Use with Lesson 7.9.

Fraction Problems

Martin brought 18 quarters to the arcade. He spent $\frac{1}{2}$ of the quarters on video games and $\frac{1}{3}$ of them on skeet-ball.

1. How much money did he spend on video games?   _____

2. How much money did he spend on skeet-ball?   _____

3. How much money did he have left?   _____

4. What fraction of the *total* is that?   _____

How many inches are in the following lengths?

1 yard = 3 feet	1 foot = 12 inches

5. $\frac{1}{2}$ yard = _____ inches

6. $\frac{1}{4}$ yard = _____ inches

7. $\frac{3}{4}$ yard = _____ inches

8. $1\frac{1}{2}$ yards = _____ inches

Shade the shape to show the fraction.

9. $\frac{2}{3}$

10. $\frac{25}{30}$

11. $\frac{12}{16}$

Math Boxes 7.9

1. Complete.

a. 17 in. = _____ ft _____ in.

b. 43 in. = _____ ft _____ in.

c. 6 ft = _____ yd

d. 11 ft = _____ yd _____ ft

e. 4 yd = _____ ft

SRB 109

2. a. What city is located at 40° N latitude and 116° E longitude?

_____

b. In which country is the city located?

_____

c. On which continent is the city located?

_____

SRB 216 217

3. a. Adena drew a line segment $\frac{3}{4}$ inch long. Then she erased a $\frac{1}{2}$ inch. How long is the line segment now?

_____ inch

b. Jordana drew a line segment $\frac{1}{4}$ inch long. Then she added another $2\frac{1}{2}$ inches. How long is the line segment now?

_____ inches

SRB 53 55

4. Write an equivalent fraction, decimal, or whole number.

	Decimal	Fraction
a.	0.40	_____
b.	_____	$\frac{3}{10}$
c.	_____	$\frac{100}{100}$
d.	0.6	_____

SRB 49 59 60

5. Sari spends $\frac{1}{3}$ of the day at school. Lunch, recess, music, gym, and art make up $\frac{1}{4}$ of her total time at school. How many hours are spent at these activities?

_____ hours

SRB 53 55

6. Multiply. Use the partial-products method.

_____ = 92 * 56

SRB 17

Use with Lesson 7.9.

The ONE

Math Message

1. If the triangle below is $\frac{1}{3}$, then what is the whole—the ONE? Draw it on the grid.

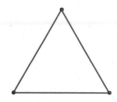

2. If $\frac{1}{4}$ of Mrs. Chin's class is 8 students, then how many students does she have altogether? _____ students

---

What Is the ONE?

Use your Geometry Template to draw the answers for Problems 3–6.

3. If ⟋_⟋ is $\frac{1}{4}$, then what is the ONE?

4. If ⬦ is $\frac{1}{2}$, then what is the ONE?

5. If ▯▯ is $\frac{2}{3}$, then what is the ONE?

6. If ⬡ is $\frac{2}{5}$, then what is the ONE?

What Is the ONE? (cont.)

Solve. If you wish, draw pictures at the bottom of the page to help you solve the problems.

7. If ⬭⬭⬭⬭⬭ is $\frac{1}{3}$, then what is the ONE?   _____ counters

8. If ⬭⬭⬭⬭ is $\frac{1}{4}$, then what is the ONE?   _____ counters

9. If 10 counters are $\frac{2}{5}$, then what is the ONE?   _____ counters

10. If 12 counters are $\frac{3}{4}$, then what is the ONE?   _____ counters

11. If $\frac{1}{5}$ of the cookies that Mrs. Jackson baked is 12, then how many cookies did she bake in all?   _____ cookies

12. In Mr. Mendez's class, $\frac{3}{4}$ of the students take music lessons. That is, 15 students take music lessons. How many students are in Mr. Mendez's class?   _____ students

Math Boxes 7.10

1. Compare.

   a. 1 day is _____ times as long as 2 hours.

   b. 6 years is _____ times as long as 4 months.

   c. 3 gallons is _____ times as much as 8 cups.

   d. 8 cm is _____ times as long as 2 mm.

   e. 1 meter is _____ times as long as 2 cm.

SRB 259

2. Multiply. Use the lattice method.

_____ = 68 * 124

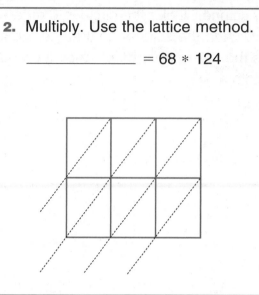

SRB 18

3. Write <, >, or = to make each sentence true.

   a. $\frac{5}{6}$ _____ $\frac{1}{6}$ **b.** $\frac{4}{10}$ _____ $\frac{4}{5}$

   c. $\frac{1}{7}$ _____ $\frac{1}{100}$ **d.** $\frac{15}{16}$ _____ $\frac{3}{4}$

   e. $\frac{7}{14}$ _____ $\frac{25}{50}$

SRB 51 52

4. Name the shaded area as a fraction and a decimal.

   a. fraction: _____

   b. decimal: _____

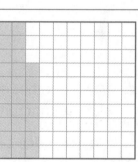

SRB 25

5. Write 5 fractions equivalent to $\frac{14}{16}$.

_____   _____

_____   _____

_____

SRB 47–49

6. Divide. Write the remainder as a fraction.

$\frac{723}{14}$ = _____

SRB 21 22 153

Making Spinners

1. Make a spinner. Color the circle in 6 different colors. Design the spinner so that the paper clip has the **same chance** of landing on each of the colors.

2. Make another spinner. Color the circle red, blue, and green so that the paper clip has

- a $\frac{1}{6}$ chance of landing on red

  and

- a $\frac{1}{3}$ chance of landing on blue.

a. What fraction of the circle did you color

red? _____   blue? _____   green? _____

b. Suppose you are about to spin the paper clip 24 times. About how many times would you expect it to land on

red? _____   blue? _____   green? _____

c. Suppose you are about to spin the paper clip 90 times. About how many times would you expect it to land on

red? _____   blue? _____   green? _____

Use with Lesson 7.11.

Counting with Fractions

Fill in the missing fractions on the number lines below.

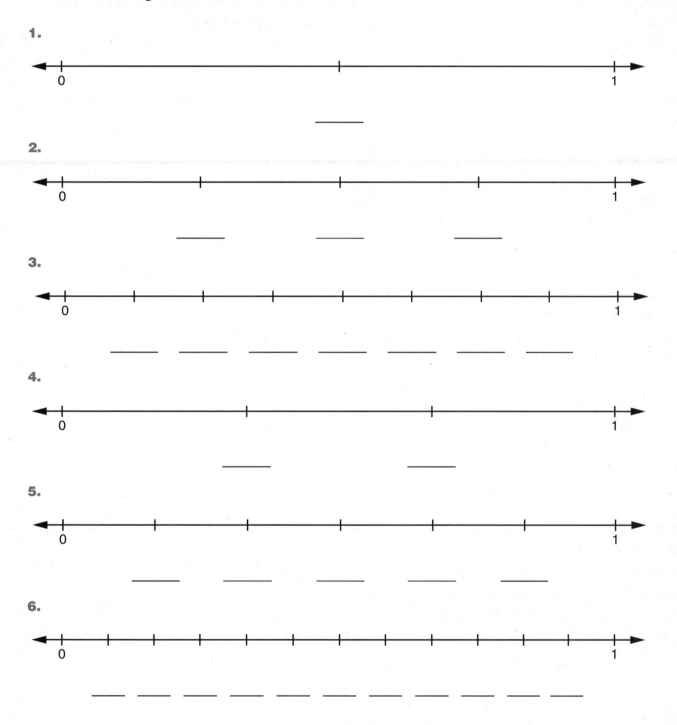

Math Boxes 7.11

1. Complete.

  a. 5 ft = _____ yd _____ ft

  b. $\frac{1}{3}$ yd = _____ in.

  c. 40 in. = _____ ft _____ in.

  d. 80 in. = _____ yd _____ in.

  e. 108 in. = _____ ft

2. a. What city is located at 60° N latitude and 110° W longitude?

    _____

  b. In which country is the city located?

    _____

  c. On which continent is the city located?

    _____

3. a. Hannah drew a line segment $1\frac{5}{8}$ inches long. Then she erased a $\frac{1}{2}$ inch. How long is the line segment now? _____ inches

  b. Joshua drew a line segment $\frac{7}{8}$ inch long. Then he added another $\frac{3}{4}$ inch. How long is the line segment now? _____ inches

4. Write an equivalent fraction, decimal, or whole number.

	Decimal	Fraction
a.	0.70	_____
b.	_____	$\frac{25}{100}$
c.	_____	$\frac{9}{9}$
d.	0.2	_____

5. According to a survey of 800 students at Martin Elementary, about $\frac{3}{4}$ of them chose pizza as their favorite food. Of those who chose pizza, $\frac{1}{2}$ liked pepperoni topping the best. How many students liked pepperoni topping the best?

    _____ students

6. Multiply. Use the partial-products method.

  71 * 38 = _____

     Use with Lesson 7.11.

Expected Spinner Results

1. If this spinner is spun 24 times, how many times do you expect it to land on each color? Fill in the table.

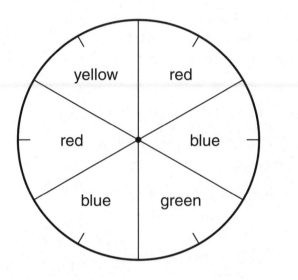

Color	Expected Number in 24 Spins
red	
blue	
yellow	
green	
Total	24

2. Explain how you made your predictions.

_____

_____

_____

_____

_____

_____

_____

_____

_____

_____

_____

Use with Lesson 7.12.

A Cube-Drop Experiment

Getting Ready

1. Follow the directions for coloring the grid on *Math Masters,* page 112. You may color the squares in any way. The colors can even form a pattern or a picture.

2. For this experiment, you are going to place your grid on the floor and hold a centimeter cube about 2 feet above the grid. Without aiming, you will let it drop onto the grid. You will then record the color of the square on which the cube finally lands.

   • If the cube does not land on the grid, the drop does not count.

   • If the cube lands on more than one color, record the color that is covered by most of the cube. If you can't tell, the toss does not count.

Making a Prediction

3. On which color is the cube *most likely* to land? _____

4. On which color is it *least likely* to land? _____

5. Suppose you were to drop the cube 100 times. How many times would you expect it to land on each color? Record your predictions below.

Predicted Results of 100 Cube Drops			
Color	**Number of Squares**	**Predicted Results**	
		Fraction	**Percent**
yellow	1	$\frac{1}{100}$	___1___ %
red	4		_____ %
green	10		_____ %
blue	35		_____ %
white	50		_____ %
Total	**100**	**1**	**100%**

A Cube-Drop Experiment (cont.)

Doing the Experiment

You and your partner will each drop a centimeter cube onto your own colored grid.

6. One partner drops the cube. The other records the color in the grid below by writing a letter in one of the squares. Drop the cube a total of 50 times. (That will fill the grid.)

Write
y for yellow,
r for red,
g for green,
b for blue, and
w for white.

7. Then trade roles. Do another 50 drops, and record the results in the other partner's journal.

8. Count the number for each color.

Write it in the "Number of Drops" column.

Check that the total is 50.

My Results for 50 Cube Drops		
Color	**Number of Drops**	**Percent**
yellow		
red		
green		
blue		
white		
Total	**50**	**100%**

9. When you have finished, fill in the percent column in the table.

Example If your cube landed on blue 15 times out of 50 drops, this is the same as 30 times out of 100 drops, or 30% of the time.

Date _____   Time _____

Fractions of Sets and Wholes

1. Circle $\frac{1}{6}$ of the triangles. Mark Xs on $\frac{2}{3}$ of the triangles.

△ △ △ △ △ △

△ △ △ △ △ △

△ △ △ △ △ △

2. a. Shade $\frac{2}{5}$ of the pentagon.

b. Shade $\frac{3}{5}$ of the pentagon.

3. There are 56 musicians in the school band: $\frac{1}{4}$ of the musicians play the flute, and $\frac{1}{8}$ play the trombone.

   a. How many musicians play the flute? _____

   b. How many musicians play the trombone? _____

4. Jennifer had 48 beanbag animals in her collection. She sold 18 of them to another collector. What fraction of her collection did she sell?

_____

5. Complete.

   a. $\frac{3}{4}$ of _____ is 90.

   b. _____ of 27 is 18.

   c. $\frac{5}{6}$ of 120 is _____.

   d. $\frac{3}{10}$ of _____ is 15.

   e. _____ of 72 is 24.

6. Fill in the missing fractions on the number line.

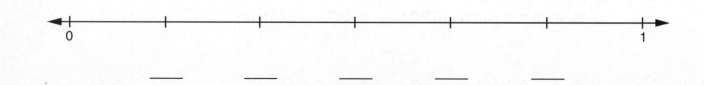

0   1

Use with Lesson 7.12.

Math Boxes 7.12

1. Compare.

   a. 1 day is _____ times as long as 6 hours.

   b. 6 years is _____ times as long as 2 months.

   c. 3 gallons is _____ times as much as 4 cups.

   d. 8 cm is _____ times as long as 5 mm.

   e. 1 meter is _____ times as long as 10 cm.

2. Multiply. Use the lattice method.

   _____ = 46 * 209

3. Write >, <, or = to make each sentence true.

   a. $\dfrac{3}{8}$ _____ $\dfrac{7}{8}$ **b.** $\dfrac{5}{12}$ _____ $\dfrac{5}{6}$

   c. $\dfrac{1}{4}$ _____ $\dfrac{1}{15}$ **d.** $\dfrac{500}{1,000}$ _____ $\dfrac{8}{16}$

   e. $\dfrac{6}{7}$ _____ $\dfrac{19}{20}$

4. Name the shaded area as a fraction and a decimal.

   a. fraction: _____

   b. decimal: _____

5. Write 5 fractions equivalent to $\dfrac{1}{6}$.

   _____   _____

   _____   _____

   _____

6. Divide. Write the remainder as a fraction.

   $\dfrac{769}{15}$ = _____

Time to Reflect

1. Suppose you had to explain to a first grader how to read the fraction $\frac{1}{6}$. What would you say?

_____

_____

_____

_____

_____

_____

2. How much does *chance* play a part in your life? Give at least two examples to support your answer.

_____

_____

_____

_____

_____

_____

3. When discussing fractions, why is it so important to know the value of the whole, or ONE? Give an example to support your answer.

_____

_____

_____

_____

_____

_____

Use with Lesson 7.13.

Math Boxes 7.13

1. Measure the length and width of your desk to the nearest half-inch. Find its perimeter.

a. Length = _____ inches

b. Width = _____ inches

c. Perimeter = _____ inches

2. Find the area of the figure.

□ = 1 square unit

Area = _____ square units

3. If 1 centimeter on a map represents 20 kilometers, then

a. 8 cm represent _____ km.

b. 3.5 cm represent _____ km.

c. _____ cm represent 30 km.

d. _____ cm represent 50 km.

e. _____ cm represents 10 km.

4. Tell if each of these is closest to 1 inch, 1 foot, or 1 yard.

a. the width of a door _____

b. the width of your ankle _____

c. the length of your little finger

_____

d. the length of your forearm _____

5. Complete.

a. 26 in. = _____ ft _____ in.

b. 57 in. = _____ ft _____ in.

c. 9 ft = _____ yd

d. 16 ft = _____ yd _____ ft

e. 8 yd = _____ ft

6. Compare.

a. 1 day is _____ times as long as 12 hours.

b. 3 years is _____ times as long as 6 months.

c. 3 gallons is _____ times as much as 2 cups.

d. 12 cm is _____ times as long as 2 mm.

e. 1 meter is _____ times as long as 20 cm.

1. A store is giving a 50% discount on all merchandise. Find the discounted prices.

Regular price	Discounted price
$26.00	_____
$0.48	_____
$140.60	_____
$64.24	_____

SRB
36 37

2. Add or subtract.

a. $\frac{5}{8} + \frac{1}{8} =$ _____

b. $\frac{1}{4} + \frac{1}{8} =$ _____

c. _____ $= \frac{11}{12} - \frac{5}{12}$

d. _____ $= \frac{5}{6} - \frac{1}{3}$

SRB
53 55

3. A group of fourth graders were asked how many minutes they spend studying at home per week. Here are the responses from ten students:

130, 45, 240, 35, 160, 185, 120, 20, 55, 160

a. What is the mode? _____

b. What is the median? _____

SRB
65

4. Insert >, <, or = to make each number sentence true.

a. $\frac{11}{12}$ _____ $\frac{19}{20}$

b. $\frac{1}{4}$ _____ $\frac{1}{9}$

c. $\frac{4}{9}$ _____ $\frac{12}{27}$

d. $\frac{10}{12}$ _____ $\frac{30}{36}$

e. $\frac{7}{2}$ _____ $\frac{21}{6}$

SRB
51 52

5. a. Use your Geometry Template to make an equilateral triangle.

b. Measure one of the angles with your protractor. Record the measure.

_____ °

SRB
122 123
84

6. If you spin the spinner below 100 times, how many times would you expect it to

land on red? _____

On black? _____

On white? _____

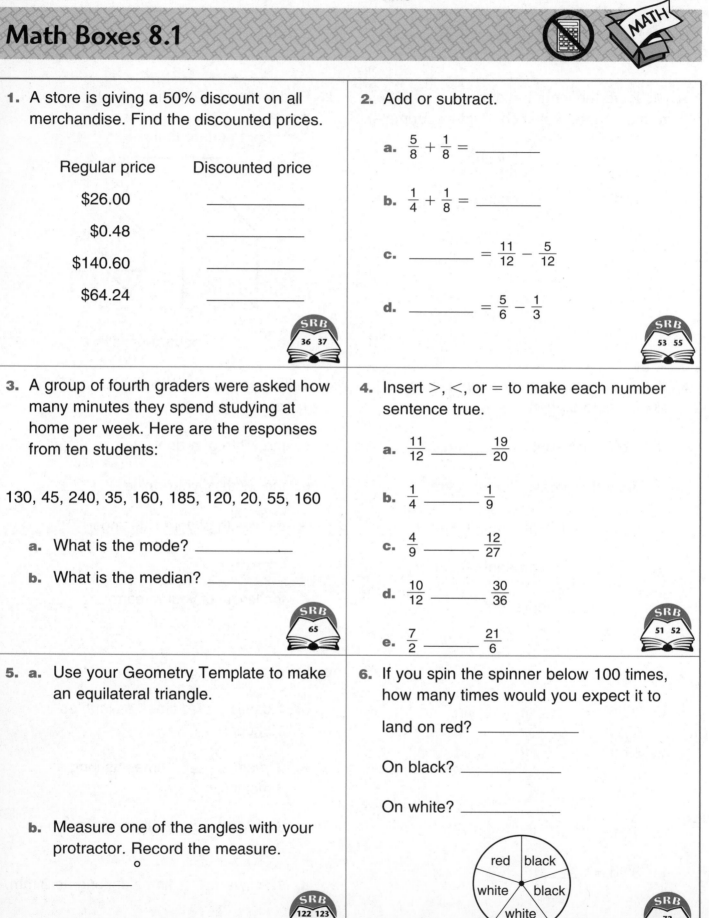

SRB
72

Use with Lesson 8.1.

Kitchen Layouts and Kitchen Efficiency

Here are four common ways to arrange the appliances in a kitchen:

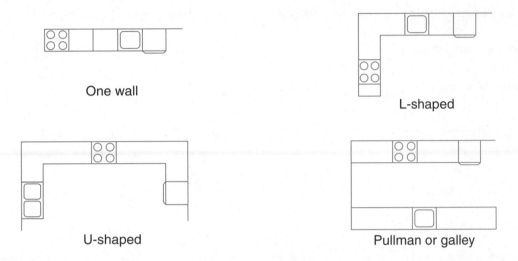

One wall

L-shaped

U-shaped

Pullman or galley

Pullman kitchens are usually found on passenger trains. **Galleys** are the kitchens on boats and airplanes. The kitchen areas on trains, boats, and airplanes are small. The cooking area is usually lined up against a single wall (a one-wall kitchen) or against two walls with a corridor between them (a Pullman or galley kitchen).

• What kind of kitchen layout do you have in your home? Circle one.

   One wall   L-shaped   U-shaped   Pullman or galley

Kitchen efficiency experts are people who study the ways we use our kitchens. They carry out **time-and-motion** studies to find how long it takes to do some kitchen tasks and how much a person has to move about in order to do them. They want to find the best ways to arrange the stove, the sink, and the refrigerator. In an efficient kitchen, a person should have to do very little walking to move from one appliance to another. However, the appliances should not be too close to each other, because the person would feel cramped.

A bird's-eye sketch is often drawn to see how well the appliances in a kitchen are arranged. The stove, the sink, and the refrigerator are connected with line segments as shown below. These segments form a triangle called a **work triangle.** The work triangle shows the distance between pairs of appliances.

Work triangle

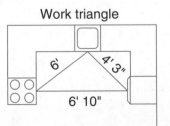

Layout of My Kitchen

1. Copy the distances between your appliances from *Math Masters*, page 322.

   Between stove and refrigerator   About _____ feet _____ inches

   Between refrigerator and sink   About _____ feet _____ inches

   Between sink and stove   About _____ feet _____ inches

2. Cut out the sketch of your kitchen from *Math Masters*, page 323 and tape it in the space below.

Use with Lesson 8.1.

How Efficient Is My Kitchen?

Answer the questions below to see how well the appliances in your kitchen are arranged.

1. With a straightedge, draw a triangle connecting the appliances in your sketch on the facing page. Write the distances between the appliances on the sides of your triangle. This triangle is called a **work triangle.**

2. Find the **perimeter** of your work triangle. Show your work.

> _____ feet _____ inches
>
> _____ feet _____ inches
>
> + _____ feet _____ inches
> _____

The perimeter is about _____ feet _____ inches.

That's close to _____ feet.

3. Kitchen efficiency experts recommend the following distances between appliances:

    Between stove and refrigerator: 4 feet to 9 feet

    Between refrigerator and sink: 4 feet to 7 feet

    Between sink and stove: 4 feet to 6 feet

    Does your kitchen meet these recommendations? _____

4. How many students reported their work triangle perimeters? _____ students

    The minimum perimeter is about _____ feet.

    The maximum perimeter is about _____ feet.

    The mode of the perimeters is about _____ feet.

    The median perimeter is about _____ feet.

Work Triangles

1. **a.** Below, draw a work triangle that meets all of the following conditions:

    • The perimeter is 21 feet.

    • The length of each side is a whole number of feet.

    • The length of each side is in the recommended range:

Between stove and refrigerator:	4 feet to 9 feet
Between refrigerator and sink:	4 feet to 7 feet
Between sink and stove:	4 feet to 6 feet

    b. Write the distances on the sides of your triangle.

    c. Label each vertex (corner) of the triangle as *stove, sink,* or *refrigerator.*

2. Below, draw a different work triangle that meets the same conditions listed above in Problem 1.

Use with Lesson 8.1.

Math Boxes 8.2

1. Shade more than $\frac{2}{100}$ but less than $\frac{1}{10}$ of the grid.

SRB
25

2. Write each number in exponential notation.

a. 100 = _____

b. 10,000 = _____

c. 1,000,000 = _____

d. 1,000 = _____

SRB
5

3. Circle the number that is closest to the product of 510 and 18.

100

1,000

10,000

100,000

SRB
155 156

4. Write an equivalent fraction, decimal, or whole number.

	Decimal	Fraction
a.	0.8	_____
b.	_____	$\frac{65}{100}$
c.	_____	$\frac{15}{15}$
d.	0.90	_____

SRB
48 59

5. Measure the sides of the figure to the nearest centimeter. Then find its perimeter.

_____ cm

_____ cm

_____ cm

_____ cm

_____ cm

Perimeter = _____ cm

SRB
108

6. If you tossed a coin onto the grid below, about what fraction of the time would you expect it to land on R?

_____

R	O	P	E
O	P	E	R
P	E	R	O
E	R	O	P

SRB
43 72

A Floor Plan of My Classroom

When architects design a room or house, they usually make two drawings. The first drawing is called a **rough floor plan.** It is not carefully drawn. But the rough floor plan includes all of the information that is needed to make an accurate drawing. The second drawing is called a **scale drawing.** It is drawn on a grid and is very accurate.

Rough floor plan for a bedroom

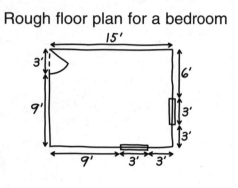

Scale drawing for a bedroom
(1 grid length represents 1 foot.)

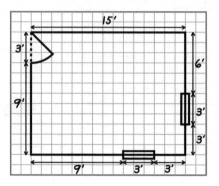

1. What information do you need to draw a rough floor plan?

_____

_____

_____

Architects use these symbols to show windows and doors:

window   door opening to left   door opening to right

2. Make a rough sketch of the outline of your classroom.

A Floor Plan of My Classroom (cont.)

3. Make a scale drawing of your classroom. Scale: _____ inch represents _____ foot.

Each side of a small square in the grid below is $\frac{1}{4}$ inch long.

For use in Lesson 8.3: The area of my classroom is about _____ square feet.

Perimeter

1. Measure the sides of the figure to the nearest centimeter. Find its perimeter.

Perimeter = _____ centimeters

2. Measure the sides of the figure to the nearest $\frac{1}{4}$ inch. Find its perimeter.

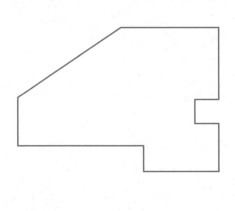

Perimeter = _____ inches

---

3. Draw a rectangle *BLUE* whose perimeter is 16 centimeters.

4. Draw a different rectangle *FARM* whose perimeter is also 16 centimeters.

---

5. Calculate the perimeter of the triangle.

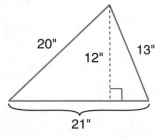

Perimeter = _____ inches

6. Calculate the perimeter of the polygon.

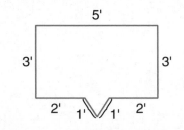

Perimeter = _____ feet

Use with Lesson 8.2.

Areas of Polygons

Find the area of each polygon.

1 cm^2
(Each side is 1 cm long.)

1.

Area = _____ cm^2

2.

Area = _____ cm^2

3.

Area = _____ cm^2

4.

Area = _____ cm^2

5.

Area = _____ cm^2

6.

Area = _____ cm^2

7.

Area = _____ cm^2

Probability

1. For this spinner, about what fraction of the spins would you expect to land on blue?

   _____

   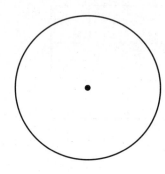

2. Design a spinner such that the probability of landing on red is $\frac{3}{8}$, on blue is $\frac{1}{4}$, on yellow is $\frac{1}{8}$, and on green is $\frac{1}{4}$.

3. If you throw a 6-sided die 84 times, how many times would you expect 4 to come up?

   _____

4. If you tossed a coin onto the grid below, what fraction of the tosses would you expect it to land on M?

   _____

   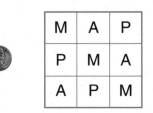

5. If you throw a 6-sided die 96 times, how many times would you expect it to land on an odd number less than 5?

   _____

6. You flip a coin 9 times, and each time the coin lands on heads. If you flip the same coin one more time, is it more likely to come up heads than tails or just as likely to come up heads as tails? Explain.

   _____

   _____

   _____

   _____

Use with Lesson 8.3.

Math Boxes 8.3

1. A store is giving a 50% discount on all merchandise. Find the discounted prices.

Regular price	Discounted price
$22.00	_____
$0.80	_____
$124.70	_____
$24.68	_____

2. Add or subtract.

a. $\frac{3}{10} + \frac{1}{10} =$ _____

b. $\frac{2}{3} + \frac{1}{6} =$ _____

c. $\frac{7}{9} - \frac{4}{9} =$ _____

d. $\frac{7}{8} - \frac{3}{4} =$ _____

3. Use the set of numbers 8, 20, 17, 16, 5, 15, and 9 to answer the questions.

What is the

a. maximum? _____

b. minimum? _____

c. range? _____

d. median? _____

4. Insert $>$, $<$, or $=$ to make each number sentence true.

a. $\frac{2}{3}$ _____ $\frac{4}{3}$

b. $\frac{4}{8}$ _____ $\frac{7}{15}$

c. $\frac{1}{15}$ _____ $\frac{1}{6}$

d. $\frac{12}{18}$ _____ $\frac{4}{6}$

e. $\frac{7}{8}$ _____ $\frac{49}{50}$

5. a. Use your Geometry Template to make a regular hexagon.

b. Measure one of the angles with your protractor. Record the measure.

_____ °

6. If you spin the spinner below 800 times, how many times would you expect it to land on red? _____

On black? _____

On white? _____

On blue? _____

What Is the Total Area of My Skin?

Follow your teacher's directions to complete this page.

1. There are _____ square inches in 1 square foot.

2. My guess is that the total
   area of my skin is about _____ square feet.

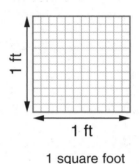

1 square foot

> *Rule of Thumb:* The total area of your skin is about
> 100 times the area of the outline of your hand.

Follow these steps to estimate the total area of your skin:

- Ask your partner to trace the outline of your hand on the grid on page 243.

- Estimate the area of the outline of your hand by counting squares on the grid.
  Record your estimate in Problem 3 below.

- Use the rule of thumb to estimate the total area of your skin (area of skin =
  100 * area of hand). Record your estimate in Problem 4 below.

3. I estimate that the area of the outline of my hand is about _____ square inches.

4. I estimate that the total area of my skin is about _____ square inches.

5. I estimate that the total area of my skin is about _____ square feet.

6. a. There are _____ square feet
      in 1 square yard.

   b. I estimate that the total area of my

      skin is about _____ square yards.

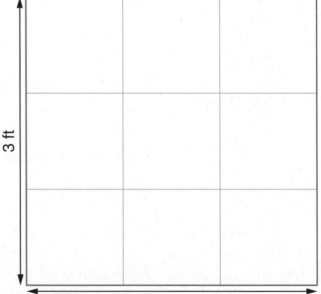

1 square yard

What Is the Total Area of My Skin? (cont.)

Ask your partner to trace the outline of your hand onto the grid below.
Keep your 4 fingers and thumb together.

Each grid square has 1-inch sides and an area of 1 square inch.

Math Boxes 8.4

1. Shade more than $\frac{18}{100}$ but less than $\frac{3}{10}$ of the grid.

2. Write each number in exponential notation.

    a. 100,000 = _____

    b. 10 = _____

    c. 10,000,000 = _____

    d. 1,000,000,000 = _____

3. Circle the number that is closest to the product of 192 and 49.

    100

    1,000

    10,000

    100,000

4. Write an equivalent fraction, decimal, or whole number.

	Decimal	Fraction
a.	0.20	_____
b.	_____	$\frac{4}{5}$
c.	_____	$\frac{0}{3}$
d.	0.1	_____

5. Measure the sides of the figure to the nearest centimeter. Then find its perimeter.

____ cm

____ cm

____ cm

____ cm

____ cm   ____ cm   ____ cm

Perimeter = _____ cm

6. If you tossed a coin onto the grid below, about what fraction of the time would you expect it to land on a vowel?

_____

R	O	P	E
O	P	E	R
P	E	R	O
E	R	O	P

     Use with Lesson 8.4.

Math Boxes 8.5

1. Divide. Write the remainder as a fraction.

5,682 / 4 = _____

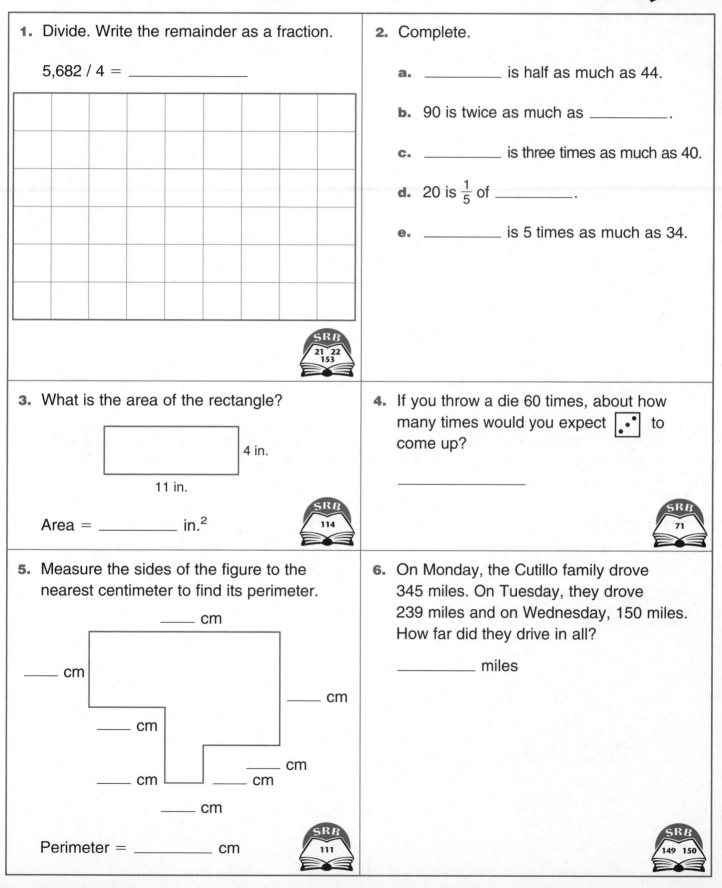

SRB
21 22
153

2. Complete.

a. _____ is half as much as 44.

b. 90 is twice as much as _____.

c. _____ is three times as much as 40.

d. 20 is $\frac{1}{5}$ of _____.

e. _____ is 5 times as much as 34.

3. What is the area of the rectangle?

4 in.

11 in.

Area = _____ in.^{2}

SRB
114

4. If you throw a die 60 times, about how many times would you expect ⚃ to come up?

_____

SRB
71

5. Measure the sides of the figure to the nearest centimeter to find its perimeter.

_____ cm

_____ cm

_____ cm

_____ cm

_____ cm

_____ cm

_____ cm

_____ cm

Perimeter = _____ cm

SRB
111

6. On Monday, the Cutillo family drove 345 miles. On Tuesday, they drove 239 miles and on Wednesday, 150 miles. How far did they drive in all?

_____ miles

SRB
149 150

Areas of Rectangles

Math Message

1. Find the area of each rectangle.

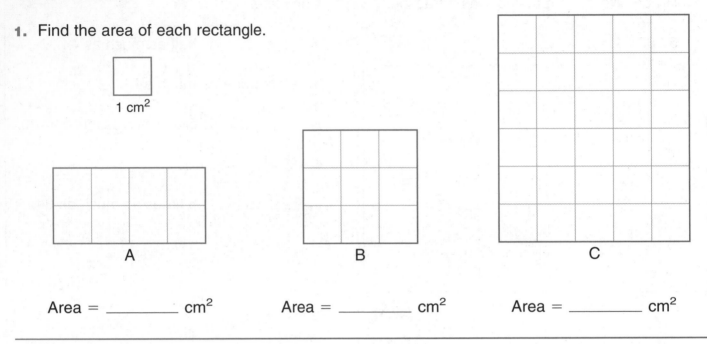

1 cm²

A

B

C

Area = _____ cm²   Area = _____ cm²   Area = _____ cm²

2. Fill in the table.

Rectangle	Number of squares per row	Number of rows	Total number of squares	Number model
A	4			
B				
C				

3. Write a formula for finding the area of a rectangle.

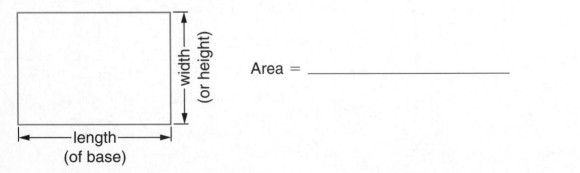

width (or height)

length (of base)

Area = _____

     Use with Lesson 8.5.

Areas of Rectangles (cont.)

4. Fill in the table at the bottom of the page.

Rectangle	Area (counting squares)	Length (of base)	Width (or height)	Area (using formula)
D	_____ cm²	_____ cm	_____ cm	_____ cm²
E	_____ cm²	_____ cm	_____ cm	_____ cm²
F	_____ cm²	_____ cm	_____ cm	_____ cm²
G	_____ cm²	_____ cm	_____ cm	_____ cm²
H	_____ cm²	_____ cm	_____ cm	_____ cm²
I	_____ cm²	_____ cm	_____ cm	_____ cm²

Scale

In each problem below, a scale and the lengths of the sides of a rectangle are given.
Make a scale drawing of each rectangle.

1. Scale: 1 cm represents 5 meters

Dimensions of rectangle:
30 meters by 35 meters

2. Scale: 1 cm represents 10 meters

Dimensions of rectangle:
40 meters by 55 meters

3. Scale: 1 inch represents 10 yards

Dimensions of rectangle:
20 yards by 30 yards

4. Scale: $\frac{1}{2}$ inch represents 10 yards

Dimensions of rectangle:
40 yards by 50 yards

Math Boxes 8.6

1. Multiply. Use your favorite method.

83 * 74 = _____

SRB
17 18

2. Add or subtract.

a. $\frac{3}{16} + \frac{7}{16}$ = _____

b. $\frac{1}{16} + \frac{1}{8}$ = _____

c. _____ = $\frac{9}{10} - \frac{3}{10}$

d. _____ = $\frac{3}{4} - \frac{3}{8}$

SRB
53 55

3. What is the area of the parallelogram?

7"

3"

Area = _____ sq in.

SRB
115

4. A jar contains

8 blue blocks,

4 red blocks,

9 orange blocks, and

4 green blocks.

You put your hand in the jar and pull out a block. About what fraction of the time would you expect to get a blue block?

_____

SRB
72

5. Dimensions for actual rectangles are given. Make scale drawings of each rectangle. Scale: 1 cm represents 20 meters.

a. Length of rectangle: 80 meters
   Width of rectangle: 30 meters

b. Length of rectangle: 90 meters
   Width of rectangle: 50 meters

SRB
125

Areas of Parallelograms

1. Cut out Parallelogram A on *Math Masters,* page 122. DO NOT CUT OUT THE
ONE BELOW. Cut it into 2 pieces so that it can be made into a rectangle.

1 cm²

Parallelogram A

Tape your rectangle in the space below.

base = _____ cm

length of base = _____ cm

height = _____ cm

width (height) = _____ cm

Area of parallelogram = _____ cm²

Area of rectangle = _____ cm²

---

2. Do the same with Parallelogram B on *Math Masters,* page 122.

Parallelogram B

Tape your rectangle in the space below.

base = _____ cm

length of base = _____ cm

height = _____ cm

width (height) = _____ cm

Area of parallelogram = _____ cm²

Area of rectangle = _____ cm²

Use with Lesson 8.6.

Areas of Parallelograms (cont.)

3. Do the same with Parallelogram C.

Parallelogram C

Tape your rectangle in the space below.

base = _____ cm

length of base = _____ cm

height = _____ cm

width (height) = _____ cm

Area of parallelogram = _____ cm^2

Area of rectangle = _____ cm^2

4. Do the same with Parallelogram D.

Parallelogram D

Tape your rectangle in the space below.

base = _____ cm

length of base = _____ cm

height = _____ cm

width (height) = _____ cm

Area of parallelogram = _____ cm^2

Area of rectangle = _____ cm^2

5. Write a formula for finding the area of a parallelogram.

_____

height

base

Areas of Parallelograms (cont.)

6. Draw a line segment to show the height of Parallelogram *DORA*.

   Use your ruler to measure the base and height.
   Then find the area.

   base = _____ cm

   height = _____ cm

   Area = _____ cm²

7. Draw the following shapes on the grid below:

   a. A rectangle whose area is 12 square centimeters.
   b. A parallelogram, not a rectangle, whose area is 12 square centimeters.
   c. A different parallelogram whose area is also 12 square centimeters.

8. What is the area of:

   a. Parallelogram *ABCD*?   b. Trapezoid *EBCD*?   c. Triangle *ABE*?

   _____ cm²   _____ cm²   _____ cm²

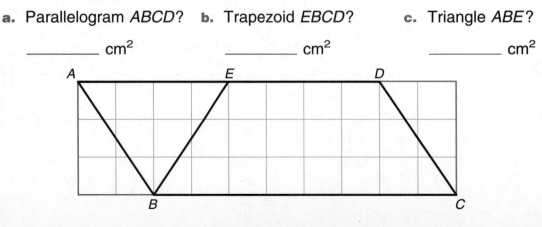

Use with Lesson 8.6.

Building a Fence

Imagine that you want to use part of your yard for a garden. The garden will be rectangular. There will be a fence around it.

You have 16 yards of material for the fence. You want the garden to have as large an area as possible. What should its dimensions be?

You can use the dot grid below to help you solve the problem. Draw different rectangles on the grid. For each rectangle, the perimeter (distance around) should be 16 yards. Find the rectangle having the largest area.

1. What is the perimeter of your garden? _____ yards

2. The garden with the largest area has a

   length of _____ yards and a width of _____ yards.

3. What is the largest possible area of your garden? _____ square yards

Areas of Triangles

1. Cut out Triangles A and B from *Math Masters,* page 123. DO NOT CUT OUT
   THE ONE BELOW. Tape the two triangles together to form a parallelogram.

   1 cm²

   Triangle A   Tape your parallelogram in the space below.

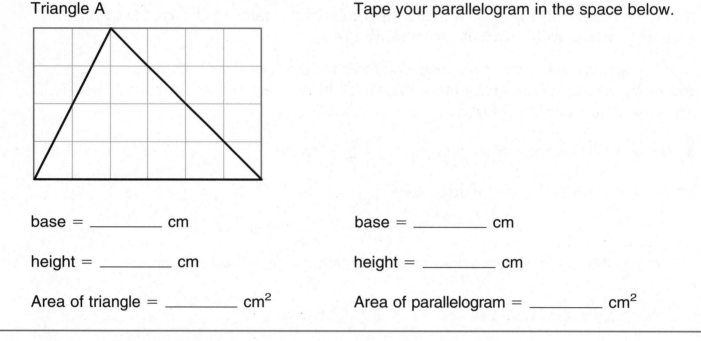

   base = _____ cm   base = _____ cm

   height = _____ cm   height = _____ cm

   Area of triangle = _____ cm²   Area of parallelogram = _____ cm²

---

2. Do the same with Triangles C and D.

   Triangle C   Tape your parallelogram in the space below.

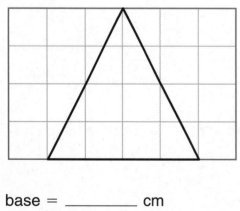

   base = _____ cm   base = _____ cm

   height = _____ cm   height = _____ cm

   Area of triangle = _____ cm²   Area of parallelogram = _____ cm²

Use with Lesson 8.7.

Areas of Triangles (cont.)

3. Do the same with Triangles E and F.

1 cm²

Triangle E

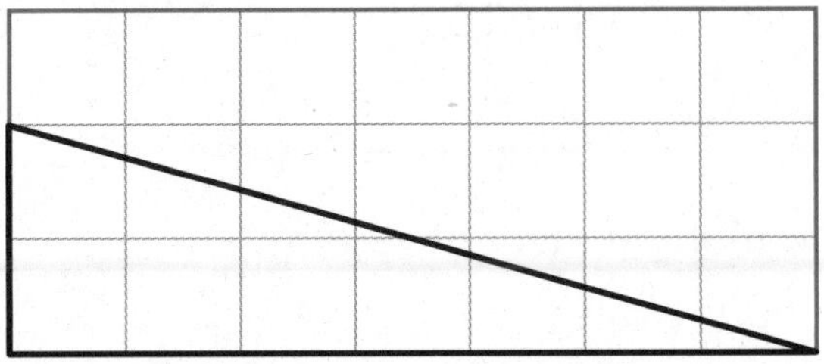

Tape your parallelogram in the space below.

base = _____ cm

height = _____ cm

Area of triangle = _____ cm²

base = _____ cm

height = _____ cm

Area of parallelogram = _____ cm²

4. Do the same with Triangles G and H.

Triangle G

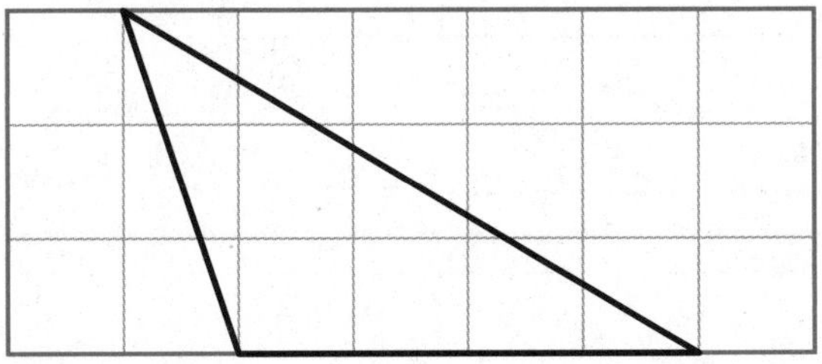

Tape your parallelogram in the space below.

base = _____ cm

height = _____ cm

Area of triangle = _____ cm²

base = _____ cm

height = _____ cm

Area of parallelogram = _____ cm²

5. Write a formula for finding
the area of a triangle.

_____

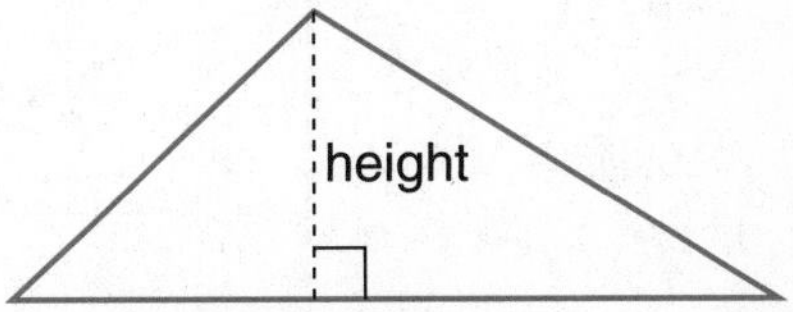

height

length of base

Areas of Triangles (cont.)

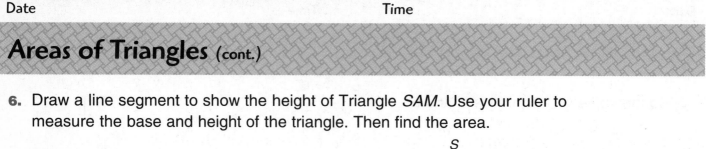

6. Draw a line segment to show the height of Triangle *SAM*. Use your ruler to measure the base and height of the triangle. Then find the area.

   base = _____ cm

   height = _____ cm

   Area = _____ cm²

7. Draw three *different* triangles on the grid below. Each triangle must have an area of 3 square centimeters. One triangle should have a right angle.

8. Which has the larger area—the star or the square? _____ Explain your answer.

   _____

   _____

   _____

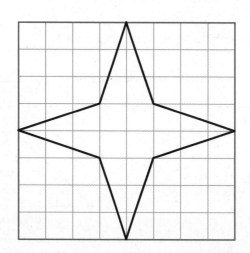
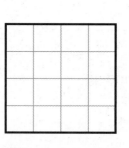

Use with Lesson 8.7.

Perimeter and Area

Use a straightedge to draw each figure.

1. Draw a parallelogram that has at least one right angle and whose area is 8 square centimeters. Find its perimeter.

Perimeter = _____ centimeters

2. Draw a parallelogram that is not a rectangle, whose height is the same length as its base, and whose area is 9 square centimeters.

3. Draw a trapezoid that has exactly two right angles. Find its area.

Area = _____ square centimeters

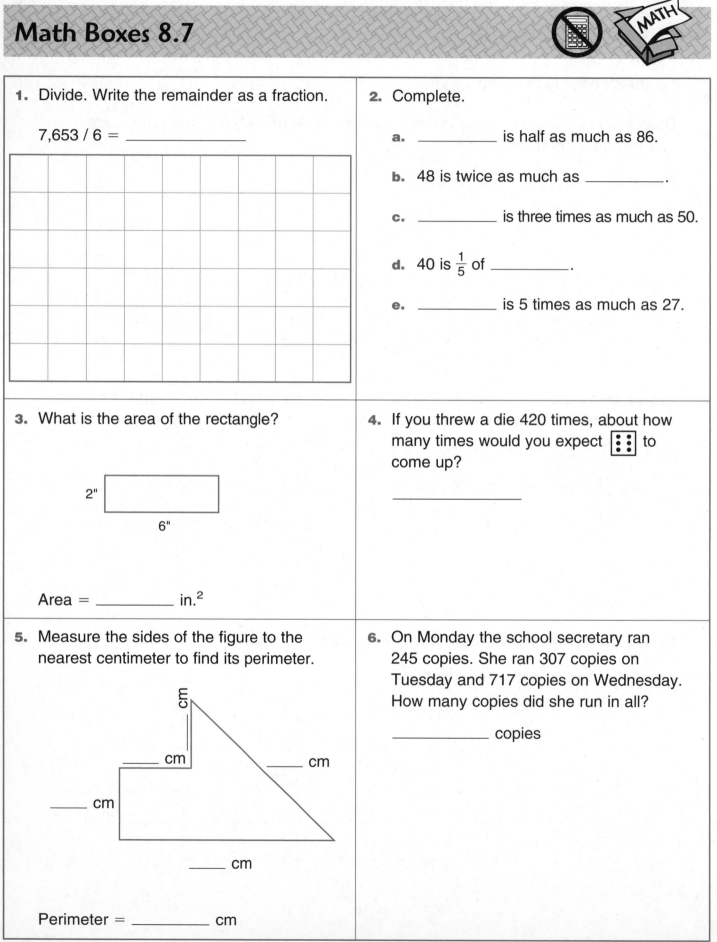

1. Divide. Write the remainder as a fraction.

7,653 / 6 = _____

2. Complete.

a. _____ is half as much as 86.

b. 48 is twice as much as _____.

c. _____ is three times as much as 50.

d. 40 is $\frac{1}{5}$ of _____.

e. _____ is 5 times as much as 27.

3. What is the area of the rectangle?

2" [6"]

Area = _____ in.²

4. If you threw a die 420 times, about how many times would you expect [die] to come up?

_____

5. Measure the sides of the figure to the nearest centimeter to find its perimeter.

_____ cm
_____ cm
_____ cm
_____ cm
_____ cm

Perimeter = _____ cm

6. On Monday the school secretary ran 245 copies. She ran 307 copies on Tuesday and 717 copies on Wednesday. How many copies did she run in all?

_____ copies

Math Boxes 8.8

1. Multiply. Use your favorite method.

$91 * 48 =$ _____

2. Add or subtract.

a. $\frac{1}{12} + \frac{11}{12} =$ _____

b. $\frac{1}{6} + \frac{2}{3} =$ _____

c. _____ $= \frac{7}{8} - \frac{5}{8}$

d. _____ $= \frac{5}{16} - \frac{1}{8}$

3. What is the area of the parallelogram?

8"

1"

Area = _____ sq in.

4. A jar contains

27 blue blocks,

18 red blocks,

12 orange blocks, and

43 green blocks.

You put your hand in the jar and pull out a block. About what fraction of the time would you expect to get a red block?

_____

5. Dimensions for actual rectangles are given. Make scale drawings of each rectangle. Scale: 1 cm represents 1.5 meters.

a. Length of rectangle: 6 meters
   Width of rectangle: 3 meters

b. Length of rectangle: 10.5 meters
   Width of rectangle: 4.5 meters

Comparing Country Areas

Brazil is the largest country in South America. Brazil's area is about 3,300,000 square miles. The area of the United States is about 3,500,000 square miles. So Brazil is nearly the same size as the United States.

Fill in the table below. This will help you to compare the areas of other countries in South America to Brazil's area. Round quotients in Part 4 to the nearest tenth.

Country	(1) Guess the number of times it would fit in the area of Brazil	(2) Area	(3) Area (rounded to the nearest 10,000)	(4) Divide the rounded areas: Brazil area ÷ country area
Ecuador	_____	109,500 mi^2	_____ mi^2	3,300,000 ÷ _____ = _____
Argentina	_____	1,068,300 mi^2	_____ mi^2	3,300,000 ÷ _____ = _____
Paraguay	_____	157,000 mi^2	_____ mi^2	3,300,000 ÷ _____ = _____
Peru	_____	496,200 mi^2	_____ mi^2	3,300,000 ÷ _____ = _____
Uruguay	_____	68,000 mi^2	_____ mi^2	3,300,000 ÷ _____ = _____
Chile	_____	292,300 mi^2	_____ mi^2	3,300,000 ÷ _____ = _____

Use with Lesson 8.8.

Comparing Country Areas (cont.)

Use your pencil to shade the number of times this country would fit into the area of Brazil.

Country Outlines

Country
Ecuador
Argentina
Paraguay
Peru
Uruguay
Chile

Area

1. What is the area of the rectangle?

1.5 cm

3.6 cm

Area = _____ sq cm

2. What is the area of the parallelogram?

$5\frac{1}{2}'$

$3\frac{1}{2}'$

Area = _____ sq ft

3. What is the area of the triangle?

6.5 m

9.2 m

Area = _____ sq m

4. Find the area of the figure.

☐ = 1 square unit

Area = _____ square units

5. Draw a rectangle with an area of 18 square centimeters and a perimeter of 22 centimeters.

Use with Lesson 8.8.

Time to Reflect

1. Which do you think is more difficult—making a rough sketch of something or making a scale drawing of something? Explain your answer.

_____

_____

_____

_____

_____

2. Think about the decorating and upkeep that your family might do to your home. When might they need to know the *area* of something?

_____

_____

_____

_____

3. Which area would it take you longer to paint—20 square yards or 20 square meters?

_____

Math Boxes 8.9

1. A store is giving a 50% discount on all merchandise. Find the discounted prices.

Regular price	Discounted price
$53.00	_____
$0.96	_____
$111.10	_____
$75.50	_____

2. Shade more than $\frac{70}{100}$ but less than $\frac{9}{10}$ of the grid.

3. Multiply. Use your favorite method.

   a. $482 * 6 =$ _____   b. $75 * 84 =$ _____   c. $36 * 58 =$ _____

4. Divide. Write the remainder as a fraction.

   a. $853 / 7 =$ _____   b. $7,342 \div 5 =$ _____   c. $\frac{385}{12} =$ _____

Use with Lesson 8.9.

Many Names for Percents

Your teacher will tell you how to fill in the percent examples.

Fill in the "100% box" for each example. Show the percent by shading the 10 by 10 square. Then write other names for the percent next to the square.

Example Last season, Duncan made 62 percent of his basketball shots.

That's ___62___ out of every 100.

100%

all of Duncan's shots

Fraction name: $\dfrac{62}{100}$

Decimal name: ___0.62___

---

1. Percent Example: _____

_____

_____

That's _____ out of every 100.

100%

Fraction name: $\dfrac{}{100}$

Decimal name: _____

---

2. Percent Example: _____

_____

_____

That's _____ out of every 100.

100%

Fraction name: $\dfrac{}{100}$

Decimal name: _____

Many Names for Percents (cont.)

Fill in the "100% box" for each example. (Problem 3 is done for you.) Show the percent by shading the 10 by 10 square. Then write other names for the percent next to the square.

3. Example 12% of the students in Marshall School are left-handed.

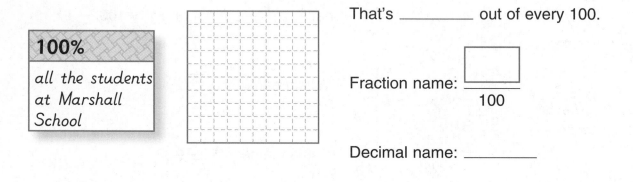

That's _____ out of every 100.

Fraction name: $\dfrac{\boxed{}}{100}$

Decimal name: _____

---

4. Example Sarah spelled 80% of the words correctly on her last test.

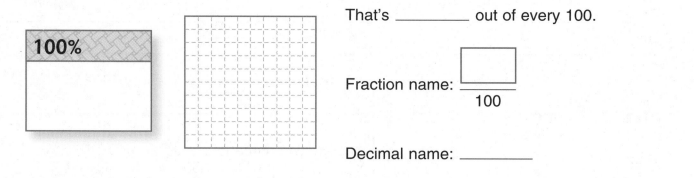

That's _____ out of every 100.

Fraction name: $\dfrac{\boxed{}}{100}$

Decimal name: _____

---

5. Example Cats sleep about 58% of the time.

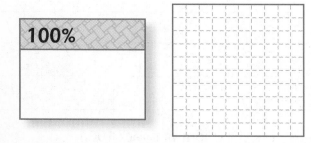 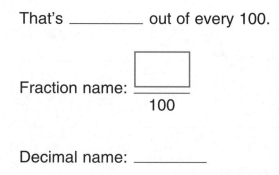

That's _____ out of every 100.

Fraction name: $\dfrac{\boxed{}}{100}$

Decimal name: _____

Use with Lesson 9.1.

Many Names for Percents (cont.)

Fill in the "100% box" for each example. Show the percent by shading the
10 by 10 square. Write other names for the percent next to the square.
Then answer the question.

6. *Example*

Sale—40% Off
Everything Must Go!

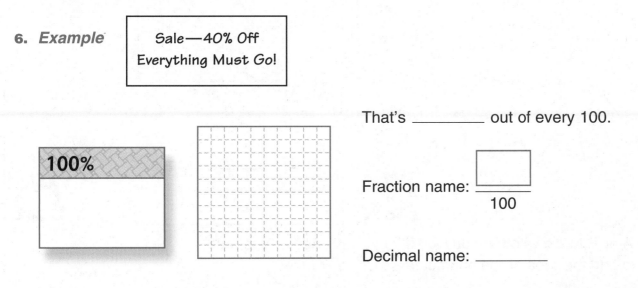

That's _____ out of every 100.

Fraction name: $\dfrac{\boxed{}}{100}$

Decimal name: _____

What would you pay for a bicycle that had been selling for $300? _____

7. *Example* A carpet store ran a TV commercial that said:
"Pay 20% when you order. Take 1 full year to pay the rest."

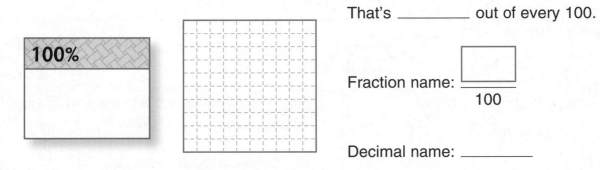

That's _____ out of every 100.

Fraction name: $\dfrac{\boxed{}}{100}$

Decimal name: _____

You order a $1,200 carpet. How much must you pay at the time
that you order it?

_____

1. Use a straightedge to draw the line of symmetry.

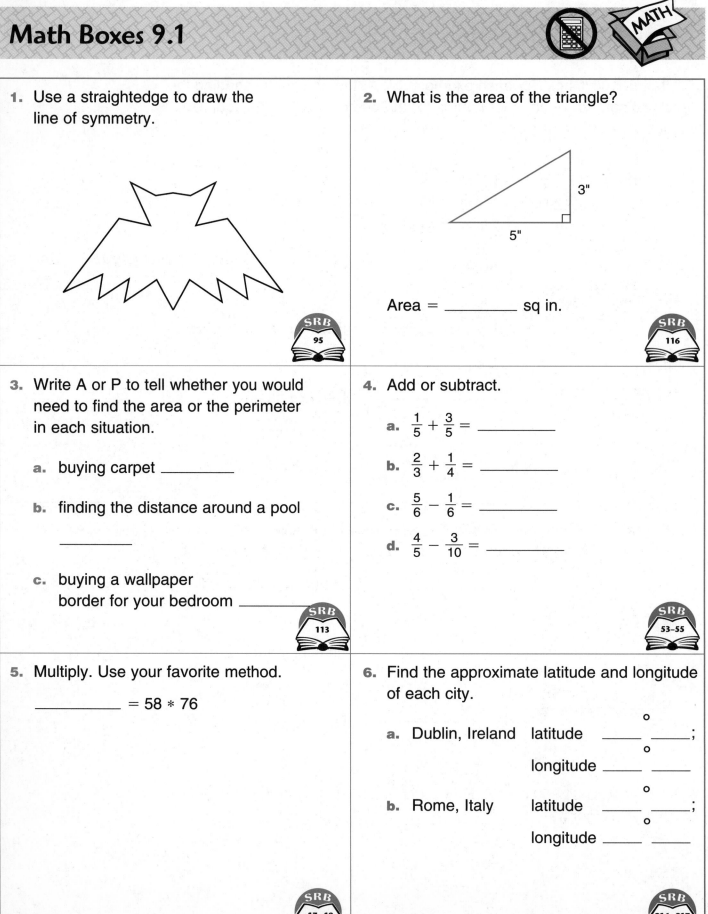

SRB 95

2. What is the area of the triangle?

3"

5"

Area = _____ sq in.

SRB 116

3. Write A or P to tell whether you would need to find the area or the perimeter in each situation.

a. buying carpet _____

b. finding the distance around a pool

_____

c. buying a wallpaper border for your bedroom _____

SRB 113

4. Add or subtract.

a. $\frac{1}{5} + \frac{3}{5} =$ _____

b. $\frac{2}{3} + \frac{1}{4} =$ _____

c. $\frac{5}{6} - \frac{1}{6} =$ _____

d. $\frac{4}{5} - \frac{3}{10} =$ _____

SRB 53–55

5. Multiply. Use your favorite method.

_____ = 58 * 76

SRB 17 18

6. Find the approximate latitude and longitude of each city.

a. Dublin, Ireland   latitude ___°___;

longitude ___°___

b. Rome, Italy   latitude ___°___;

longitude ___°___

SRB 216 217

"Percent-of" Number Stories

Alfred, Nadine, Kyla, and Jackson each took the same math test. There were 20 problems on the test.

100%
20–problem test

1. Alfred missed $\frac{1}{2}$ of the problems. He missed **0.50** of the problems. That's **50%** of the problems.

   How many problems did he miss? _____ problems

   $\frac{1}{2}$ of 20 = _____

   50% of 20 = _____

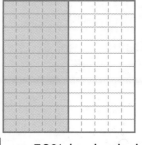

$\frac{1}{2}$, or 50% is shaded.

2. Nadine missed $\frac{1}{4}$ of the problems. She missed **0.25** of the problems. That's **25%** of the problems.

   How many problems did she miss? _____ problems

   $\frac{1}{4}$ of 20 = _____

   25% of 20 = _____

$\frac{1}{4}$, or 25% is shaded.

3. Kyla missed $\frac{1}{10}$ of the problems. She missed **0.10** of the problems. That's **10%** of the problems.

   How many problems did she miss? _____ problems

   $\frac{1}{10}$ of 20 = _____

   10% of 20 = _____

$\frac{1}{10}$, or 10% is shaded.

4. Jackson missed $\frac{1}{5}$ of the problems. He missed **0.20** of the problems. That's **20%** of the problems.

   How many problems did he miss? _____ problems

   $\frac{1}{5}$ of 20 = _____

   20% of 20 = _____

$\frac{1}{5}$, or 20% is shaded.

Equivalent Fractions, Decimals, and Percents

Fill in the missing numbers. Problem 1 has been done for you.

100%

large square

1. Ways of showing $\frac{3}{4}$:

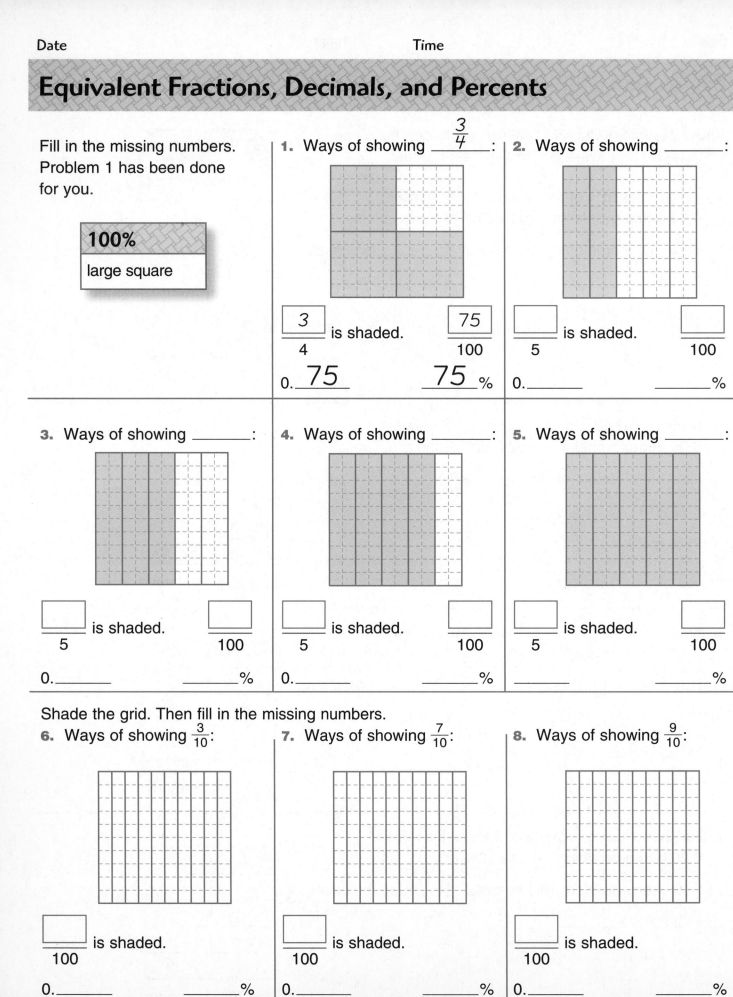

$\boxed{3}$ is shaded. $\boxed{75}$
4   100

0.$\underline{75}$ $\underline{75}$ %

2. Ways of showing _____:

$\boxed{}$ is shaded. $\boxed{}$
5   100

0._____   _____%

3. Ways of showing _____:

$\boxed{}$ is shaded. $\boxed{}$
5   100

0._____   _____%

4. Ways of showing _____:

$\boxed{}$ is shaded. $\boxed{}$
5   100

0._____   _____%

5. Ways of showing _____:

$\boxed{}$ is shaded. $\boxed{}$
5   100

_____   _____%

Shade the grid. Then fill in the missing numbers.

6. Ways of showing $\frac{3}{10}$:

$\boxed{}$ is shaded.
100

0._____   _____%

7. Ways of showing $\frac{7}{10}$:

$\boxed{}$ is shaded.
100

0._____   _____%

8. Ways of showing $\frac{9}{10}$:

$\boxed{}$ is shaded.
100

0._____   _____%

Use with Lesson 9.2.

Multiplying Whole Numbers

Multiply. Use your favorite method.

1. 9 * 78 = _____

2. 437 * 8 = _____

3. _____ = 93 * 64

4. _____ = 46 * 82

5. 27 * 534 = _____

6. 72 * 238 = _____

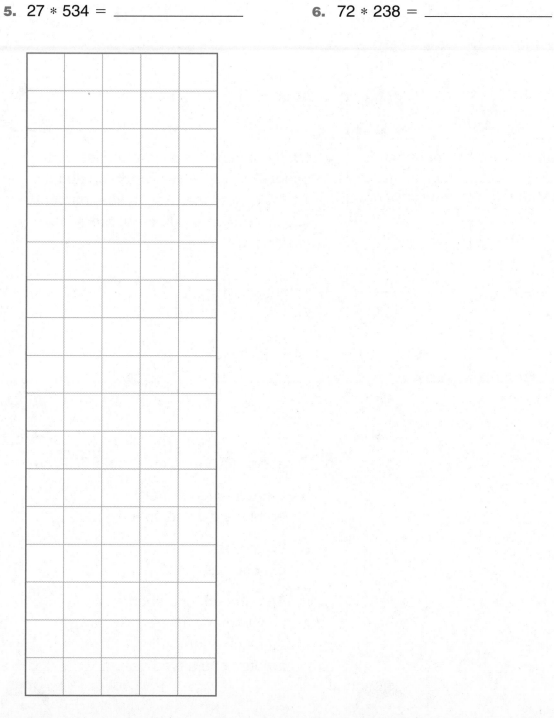

Math Boxes 9.2

1. Name the opposite of

   a. cold   _____

   b. morning   _____

   c. 5   _____

   d. −19   _____

SRB 58

2. What is the area of the rectangle?

7"

4"

Area = _____ in.^{2}

SRB 114

3. Complete the table with equivalent names.

Fraction	Decimal	Percent
$\frac{1}{5}$		
	0.80	
$\frac{3}{10}$		
		90%

SRB 59 60

4. On the Internet, the word *the* is used about 4.02% and the word *and* about 1.68% of the time. About what percent of all words on the Internet are either *the* or *and*?

_____

SRB 149 150

5. Divide.

420 ÷ 15 = _____

SRB 21 22

6. Zena earned $12. She spent $8.

   a. What fraction of her earnings did she spend? _____

   b. What fraction did she have left? _____

   c. The amount she spent is how many times as much as the amount she saved? _____

SRB 42

Use with Lesson 9.2.

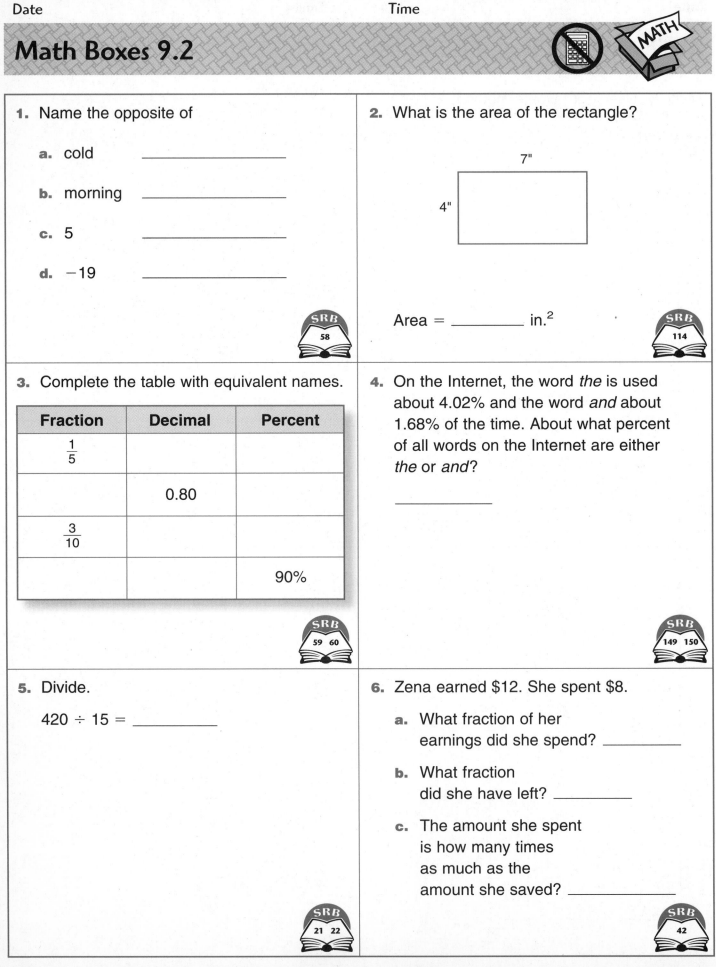

Math Boxes 9.3

1. Use a straightedge to draw the line of symmetry.

2. What is the area of the triangle?

4"

9"

Area = _____ sq in.

3. Write A or P to tell whether you would need to find the area or the perimeter in each situation.

a. buying a garden fence _____

b. finding the square footage of your bedroom _____

c. buying wallpaper for the kitchen

_____

4. Add or subtract.

a. $\frac{3}{8} + \frac{1}{8} =$ _____

b. $\frac{2}{5} + \frac{1}{10} =$ _____

c. $\frac{5}{12} - \frac{3}{12} =$ _____

d. $\frac{5}{6} - \frac{1}{3} =$ _____

5. Multiply. Use your favorite method.

_____ = 64 * 23

6. Find the approximate latitude and longitude of each city.

a. Calcutta, India   latitude _____ °_____;

longitude _____ °_____

b. Seoul, Korea   latitude _____ °_____;

longitude _____ °_____

Discount Number Stories

1. A store is offering a **discount** of 10% on all items. This means that you save
$\frac{1}{10}$ of the **regular price.** Find the sale price of each item below. The **sale price**
is the amount you pay after subtracting the discount from the regular price.

Item	Regular Price	Discount (10% of regular price)	Sale Price (Subtract: regular price – discount)
CD player	$140	$14	
Giant screen TV	$1,200		
Radio	$80		
Cassette player		$3	

2. An airline offers a 25% discount on the regular airfare for tickets purchased
at least 1 month in advance. Find the sale price of each ticket below.

Regular Airfare	Discount (25% of regular airfare)	Sale Price (Subtract: regular airfare – discount)
$400	$100	
$240		
	$75	

3. A swing set can be purchased at a 30% discount if it is ordered before April 1.
On April 1, the regular price of $400 will be charged. If you order the swing set
before April 1,

a. how much will you save? _____

b. how much will you pay? _____

Challenge

4. You can pay for a refrigerator by making 12 payments of $50 each. But you can
save 25% if you pay for it all at once.

How much will the refrigerator cost if you pay for it all at once? _____

Math Boxes 9.4

1. Name the opposite of

   a. happy   _____

   b. pull   _____

   c. -3.6 _____

   d. 10.1   _____

2. What is the area of the rectangle?

6"

2"

Area = _____ in.^{2}

3. Complete the table with equivalent names.

Fraction	Decimal	Percent
$\frac{3}{4}$		
	0.6	
$\frac{1}{10}$		
		50%

4. On the Internet, the word *a* is used about 1.96% and the word *it* about 0.81% of the time. About what percent of all words on the Internet are either *a* or *it*?

_____

5. Divide.

264 ÷ 12 = _____

6. Three friends cut a pizza into 12 equal slices and share the pizza equally.

   a. What fraction of the pizza does each get? _____

   b. How many slices does each friend get? _____ slices

Math Boxes 9.5

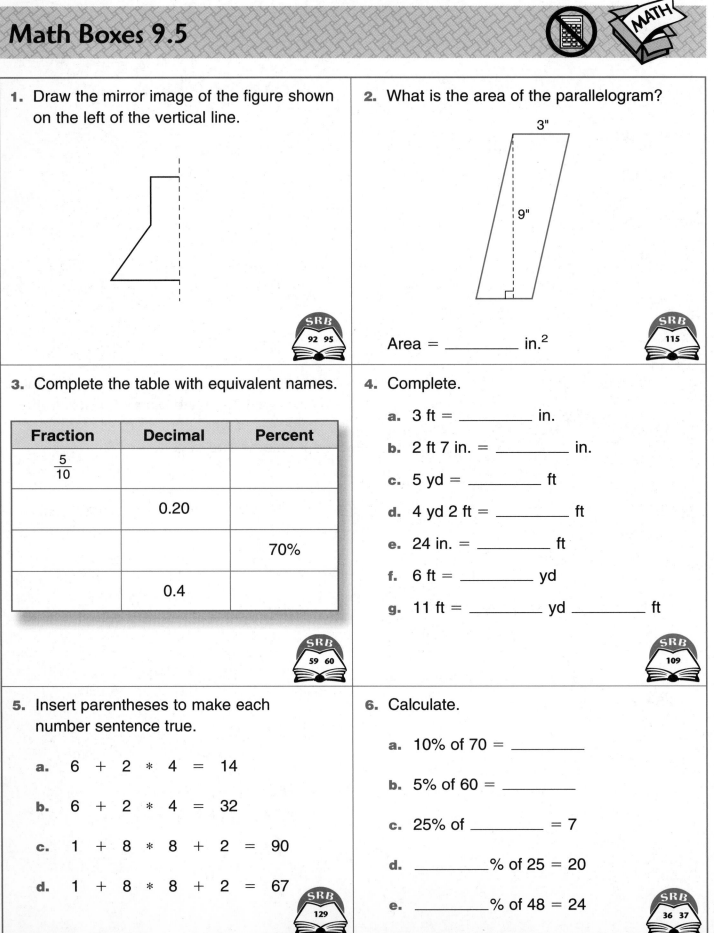

1. Draw the mirror image of the figure shown on the left of the vertical line.

SRB
92 95

2. What is the area of the parallelogram?

3"

9"

Area = _____ in.²

SRB
115

3. Complete the table with equivalent names.

Fraction	Decimal	Percent
$\frac{5}{10}$		
	0.20	
		70%
	0.4	

SRB
59 60

4. Complete.

a. 3 ft = _____ in.

b. 2 ft 7 in. = _____ in.

c. 5 yd = _____ ft

d. 4 yd 2 ft = _____ ft

e. 24 in. = _____ ft

f. 6 ft = _____ yd

g. 11 ft = _____ yd _____ ft

SRB
109

5. Insert parentheses to make each number sentence true.

a. 6 + 2 * 4 = 14

b. 6 + 2 * 4 = 32

c. 1 + 8 * 8 + 2 = 90

d. 1 + 8 * 8 + 2 = 67

SRB
129

6. Calculate.

a. 10% of 70 = _____

b. 5% of 60 = _____

c. 25% of _____ = 7

d. _____% of 25 = 20

e. _____% of 48 = 24

SRB
36 37

Use with Lesson 9.5.

Trivia Survey Results

1. The chart below will show the results of the trivia survey for the whole class. Wait for your teacher to explain how to fill in the chart.

Class Results for the Trivia Survey

Question	Yes	No	Total	$\dfrac{\text{Yes}}{\text{Total}}$	% Yes
1. Is Monday your favorite day?					
2. Have you gone to the movies in the last month?					
3. Did you eat breakfast today?					
4. Do you keep a map in your car?					
5. Did you eat at a fast-food restaurant yesterday?					
6. Did you read a book during the last month?					
7. Are you more than 1 meter tall?					
8. Do you like liver?					

2. On the basis of the survey results, is it more likely that a person will

   a. read a book or go to a movie?

   _____

   b. eat breakfast or eat at a fast-food restaurant?

   _____

   c. like liver or like Mondays?

   _____

Dividing Whole Numbers

Solve each division problem. Write the answer as a mixed number
by writing the remainder as a fraction.

1. 93 / 5 = _____

2. 89 / 4 = _____

3. _____ = 937 / 8

4. _____ = 853 / 5

5. 532 / 23 = _____

6. 674 / 12 = _____

Math Boxes 9.6

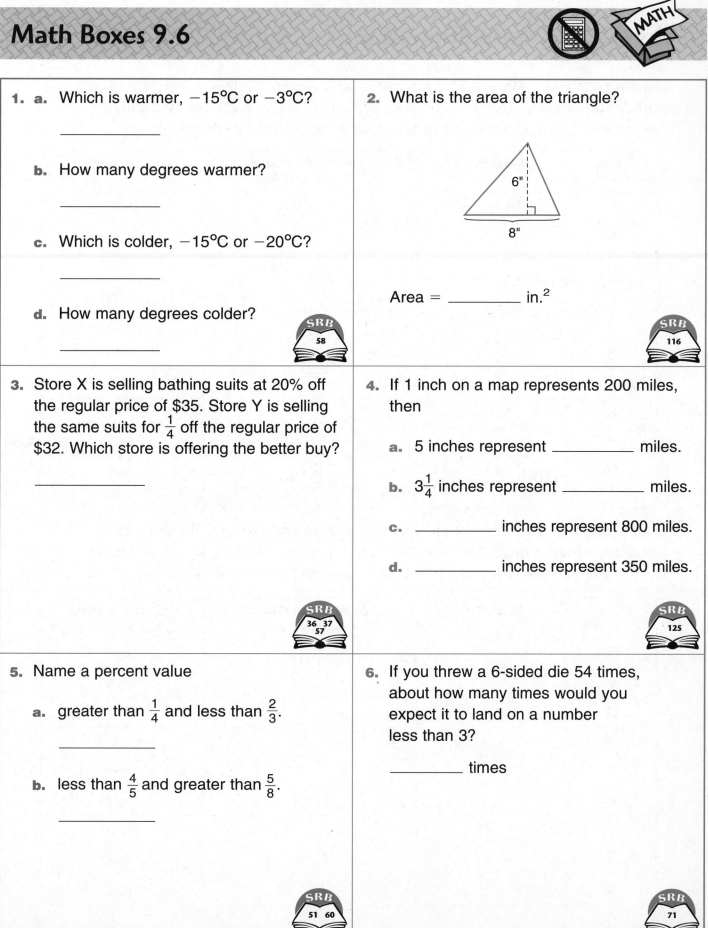

1. a. Which is warmer, −15°C or −3°C?

_____

b. How many degrees warmer?

_____

c. Which is colder, −15°C or −20°C?

_____

d. How many degrees colder?

_____

SRB 58

2. What is the area of the triangle?

6"

8"

Area = _____ in.²

SRB 116

3. Store X is selling bathing suits at 20% off the regular price of $35. Store Y is selling the same suits for $\frac{1}{4}$ off the regular price of $32. Which store is offering the better buy?

_____

SRB 36 37 57

4. If 1 inch on a map represents 200 miles, then

a. 5 inches represent _____ miles.

b. $3\frac{1}{4}$ inches represent _____ miles.

c. _____ inches represent 800 miles.

d. _____ inches represent 350 miles.

SRB 125

5. Name a percent value

a. greater than $\frac{1}{4}$ and less than $\frac{2}{3}$.

_____

b. less than $\frac{4}{5}$ and greater than $\frac{5}{8}$.

_____

SRB 51 60

6. If you threw a 6-sided die 54 times, about how many times would you expect it to land on a number less than 3?

_____ times

SRB 71

Color-Coded Population Maps

1. List the countries in Region 4 from *smallest to largest* according to the **percent of population, ages 0–14.** Take one copy of *Math Masters,* page 142 and write a title for your first map. Color these countries using the color code shown below.

Rank	Country	Percent of Population Ages 0–14	Color Code
1	*Japan*	*15%*	blue
2			blue
3			blue
4			green
5			green
6			green
7			green
8			red
9			red
10	*Bangladesh*	*38%*	red

2. List the countries in Region 4 from *smallest to largest* according to the **percent of population that is rural.** Take another copy of *Math Masters,* page 142 and write a title for your second map. Color these countries using the color code shown below.

Rank	Country	Percent of Rural Population	Color Code
1			blue
2			blue
3			blue
4			green
5			green
6			green
7			green
8			red
9			red
10			red

Math Boxes 9.7

---

1. Draw the mirror image of the figure shown on the left of the vertical line.

2. What is the area of the parallelogram?

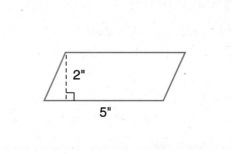

Area = _____ in.²

---

3. Complete the table with equivalent names.

Fraction	Decimal	Percent
		63%
	1.00	
$\frac{3}{5}$		
		80%

4. Complete.

a. 3 yd 2 ft = _____ ft

b. 6 yd 1 ft = _____ ft

c. 2 ft 9 in. = _____ in.

d. 25 ft = _____ yd _____ ft

e. _____ ft = 5 yd 2 ft

f. _____ in. = 2 yd

g. _____ ft _____ in. = 30 in.

---

5. Insert parentheses to make each number sentence true.

a. 3 * 5 + 6 < 3 * 10

b. 34 − 48 / 8 + 4 = 32

c. 6 * 7 + 1 < 80 / 2 + 5

d. 63 / 21 − 12 = 7

6. Calculate.

a. 10% of 50 = _____

b. 5% of 80 = _____

c. 20% of _____ = 8

d. _____% of 16 = 12

e. _____% of 24 = 6

Multiplying Decimals

Math Message

Toni has 8 nickels. Each nickel has a thickness
of 0.2 centimeter. If she puts the nickels in a stack,
what will be the height of the stack?   _____ cm

---

1. Devon measured the length of the room by pacing it off.
The length of his pace was 2.3 feet. He counted 14 paces.
How long is the room?   _____ ft

2. Lemons cost $0.35 each. How much will 25 lemons cost? $_____

3. Find the area of each rectangle below.

a. 1.5 cm ⌐─────────────────────────────⌐   Area = _____ cm²
              30 cm

b. 6 in. ⌐─────────⌐   Area = _____ in.²
            15.4 in.

4. For each problem below, the multiplication has been done correctly,
but the decimal point is missing in the answer. Correctly place the
decimal point in the answer.

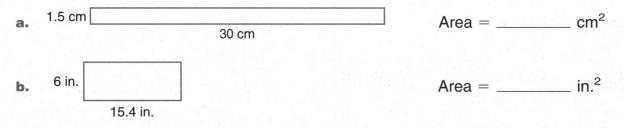

a. 23 * 7.3 = 1 6 7 9   b. 0.38 * 51 = 1 9 3 8

c. 6.91 * 82 = 5 6 6 6 2   d. 5,203 * 12.6 = 6 5 5 5 7 8

Multiplying Decimals (cont.)

For each problem:

• Estimate the product.

• Multiply the factors as though they were whole numbers.

• Use the estimate to help you place the decimal in the answer.

5. 2.7 * 45 = _____

6. 22 * 0.32 = _____

7. 0.02 * 333 = _____

8. 8 * 5.7 = _____

9. 5.08 * 27 = _____

10. 42 * 0.97 = _____

Math Boxes 9.8

1. a. Which is warmer, −7°C or −3.5°C?

   _____

   b. How many degrees warmer?

   _____

   c. Which is colder, −18°C or −9.6°C?

   _____

   d. How many degrees colder?

   _____

2. What is the area of the triangle?

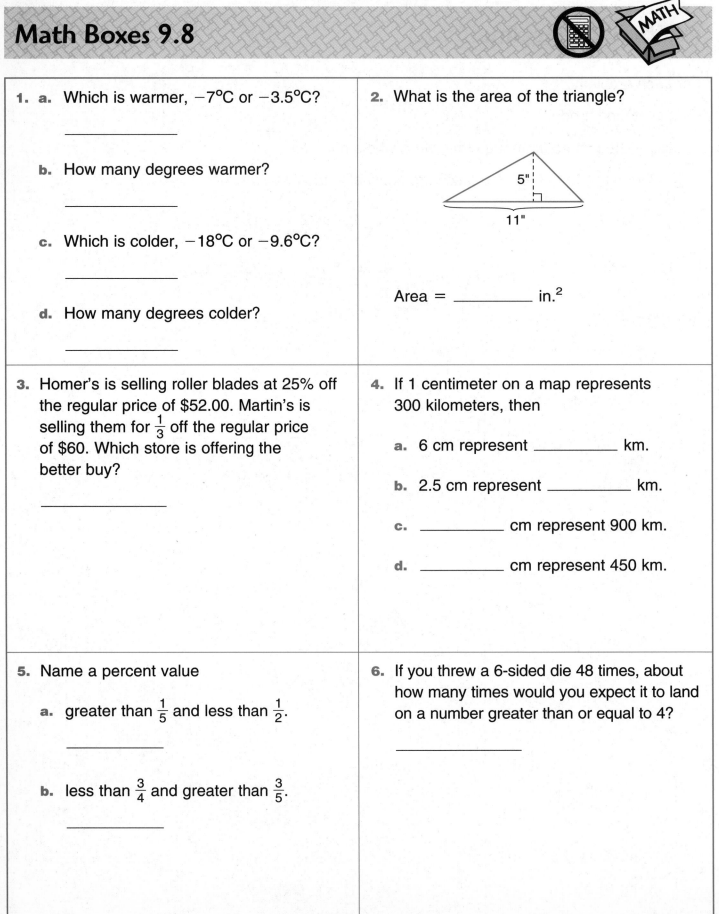

   5"

   11"

   Area = _____ in.²

3. Homer's is selling roller blades at 25% off the regular price of $52.00. Martin's is selling them for $\frac{1}{3}$ off the regular price of $60. Which store is offering the better buy?

   _____

4. If 1 centimeter on a map represents 300 kilometers, then

   a. 6 cm represent _____ km.

   b. 2.5 cm represent _____ km.

   c. _____ cm represent 900 km.

   d. _____ cm represent 450 km.

5. Name a percent value

   a. greater than $\frac{1}{5}$ and less than $\frac{1}{2}$.

   _____

   b. less than $\frac{3}{4}$ and greater than $\frac{3}{5}$.

   _____

6. If you threw a 6-sided die 48 times, about how many times would you expect it to land on a number greater than or equal to 4?

   _____

Dividing Decimals

1. Janine is building a bookshelf.
   She has a board that is 3.75 meters long.
   She wants to cut it into 5 pieces of equal length.
   What will be the length of each piece?

   _____ meter

2. Three sisters set up a lemonade stand.
   On Wednesday they made $8.46.
   If they shared the money equally,

   how much did each girl get? $_____

3. Alex and his three friends went out to lunch.
   The total bill, including tax and tip, was $42.52.
   They decided that each would pay the same amount.

   How much did each person pay? $_____

4. Victor divides a 98.4 cm piece of string into
   3 equal pieces. What is the length of each piece?

   _____ cm

For each problem below, the division has been done correctly, but the decimal point is missing in the answer. Correctly place the decimal point in the answer.

5.
$$\begin{array}{r} 1\,4\,6 \\ 3\overline{)4\,3.8} \end{array}$$

6.
$$\begin{array}{r} 8\,7\,0\,0 \\ 5\overline{)4.3\,5} \end{array}$$

7.
$$\begin{array}{r} 1\,6\,1\,5 \\ 4\overline{)6.4\,6} \end{array}$$

8.
$$\begin{array}{r} 4\,9\,8\,2\,0 \\ 6\overline{)2\,9\,8.9\,2} \end{array}$$

Dividing Decimals (cont.)

For each problem:

• Estimate the quotient.

• Divide the numbers as though they were whole numbers.

• Use the estimate to place the decimal point in the answer.

9. 89.6 / 4 = _____

10. 2.96 / 8 = _____

11. _____ = 3.65 ÷ 5

12. _____ = 9.44 / 4

13. 253.8 / 6 = _____

14. 46.8 ÷ 12 = _____

Use with Lesson 9.9.

Review: Fractions, Decimals, and Percents

1. Fill in the missing numbers in the table of equivalent fractions, decimals, and percents.

Fraction	Decimal	Percent
$\frac{4}{10}$		
	0.6	
		75%

2. Kendra set a goal of saving $50 in 8 weeks. During the first 2 weeks, she was able to save $10.

   a. What fraction of the $50 did she save in the first 2 weeks? _____

   b. What percent of the $50 did she save? _____

   c. At this rate, how long will it take her to reach her goal? _____ weeks

3. Shade 80% of the square.

   a. What fraction of the square did you shade? _____

   b. Write this fraction as a decimal. _____

   c. What percent of the square is *not* shaded? _____

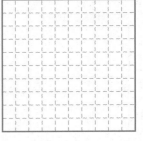

4. Tanara's new skirt was on sale at 15% off the original price. The original price of the skirt was $60.

   a. How much money did Tanara save with the discount? _____

   b. How much did she pay for the skirt? _____

5. Star Video and Vic's Video Mart sell videos at about the same regular prices. Both stores are having sales. Star Video is selling its videos at $\frac{1}{3}$ off the regular price. Vic's Video Mart is selling its videos at 25% off the regular price. Which store has the better sale? Explain your answer.

   _____

   _____

   _____

Math Boxes 9.9

1. Study the figure. Draw the other half along the vertical line of symmetry.

2. What is the area of the parallelogram?

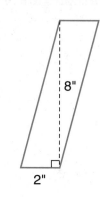

Area = _____ sq in.

3. Complete the table with equivalent names.

Fraction	Decimal	Percent
		70%
	0.75	
$\frac{3}{5}$		
		72%

4. Complete.

a. 6 yd 8 ft = _____ ft

b. 5 yd 3 ft = _____ ft

c. 4 ft 7 in. = _____ in.

d. 35 ft = _____ yd _____ ft

e. _____ ft = 7 yd 9 ft

f. _____ in. = 6 yd

g. _____ ft _____ in. = 45 in.

5. Insert parentheses to make each number sentence true.

a. 4 * 6 + 3 > 3 * 10

b. 17 − 24 / 6 + 6 = 19

c. 40 * 30 + 60 < 100 * 20

d. 56 / 7 − 3 = 14

6. Calculate.

a. 10% of 90 = _____

b. 5% of 140 = _____

c. 20% of _____ = 9

d. _____% of 30 = 24

e. _____% of 48 = 36

Use with Lesson 9.9.

Time to Reflect

1. Think of some numbers you have seen in the past week in a grocery store or in a magazine. Do you think you have seen more decimals, more percents, or more fractions? Try to explain why.

_____

_____

_____

_____

_____

2. Did *Fraction/Percent Concentration* help you learn some of the easy fraction/percent equivalencies? If not, what strategy do you think would help you learn them?

_____

_____

_____

_____

_____

_____

3. Give two examples of uses of percents.

_____

_____

_____

4. Suppose you are conducting a survey. You want to know what 9-year-olds like best about living in the country. Would you ask people over 20 years old any questions? Explain your answer.

_____

_____

_____

_____

Use with Lesson 9.10.

Math Boxes 9.10

1. Use a straightedge to draw the line of symmetry.

2. Name the opposite of

a. honest _____

b. rough _____

c. 8.4 _____

d. $-\frac{4}{5}$ _____

3. Draw the mirror image of the figure shown on the left of the vertical line.

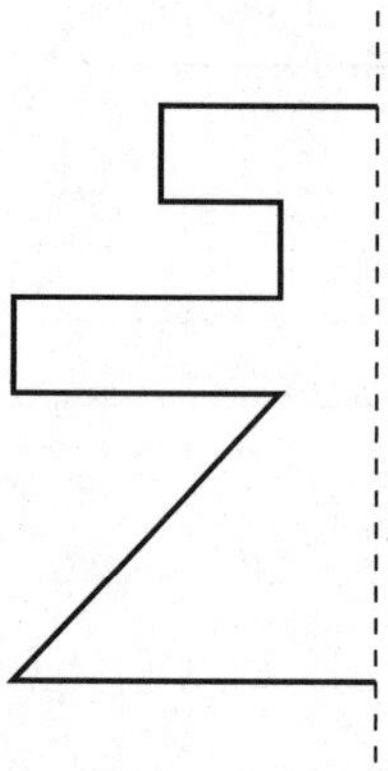

4. a. Which is warmer, $-9.4°C$ or $-11.2°C$?

_____

b. How many degrees warmer?

_____

c. Which is colder, $-19.3°C$ or $-12.8°C$?

_____

d. How many degrees colder?

_____

5. Name four numbers greater than -8 and less than -5.

_____

_____

_____

_____

6. Name four numbers less than 2 and greater than -1.

_____

_____

_____

_____

Use with Lesson 9.10.

Basic Use of a Transparent Mirror

A **transparent mirror** is shown at the right.

Notice that the mirror has a **recessed** drawing edge, along which lines are drawn. Some transparent mirrors have a drawing edge both on the top and on the bottom.

Place your transparent mirror on this page so that its drawing edge lies along line *MK* below. Then look through the transparent mirror to read the "backward" message.

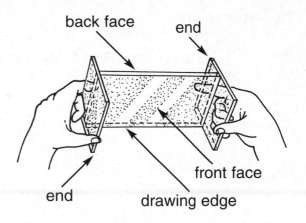

back face   end

front face

end   drawing edge

M   *K*

If you have followed the directions correctly, you are now able to read this message. Here are a few things to remember when using your transparent mirror:

- Always look into the front of the transparent mirror.
- Use your transparent mirror on flat surfaces like your desk or a tabletop. In this position, the drawing edge will be facing you.
- Use a sharp pencil when tracing along the drawing edge.
- Experiment and have fun!

Multiplying and Dividing with Decimals

Multiply. Show your work.

1. 5.7 ∗ 52 = _____   **2.** 93 ∗ 0.48 = _____   **3.** _____ = 3.85 ∗ 27

Divide. Show your work.

4. 7)‾33.6‾   **5.** 30.4 ÷ 8 = _____   **6.** _____ = 198.9 / 9

Use with Lesson 10.1.

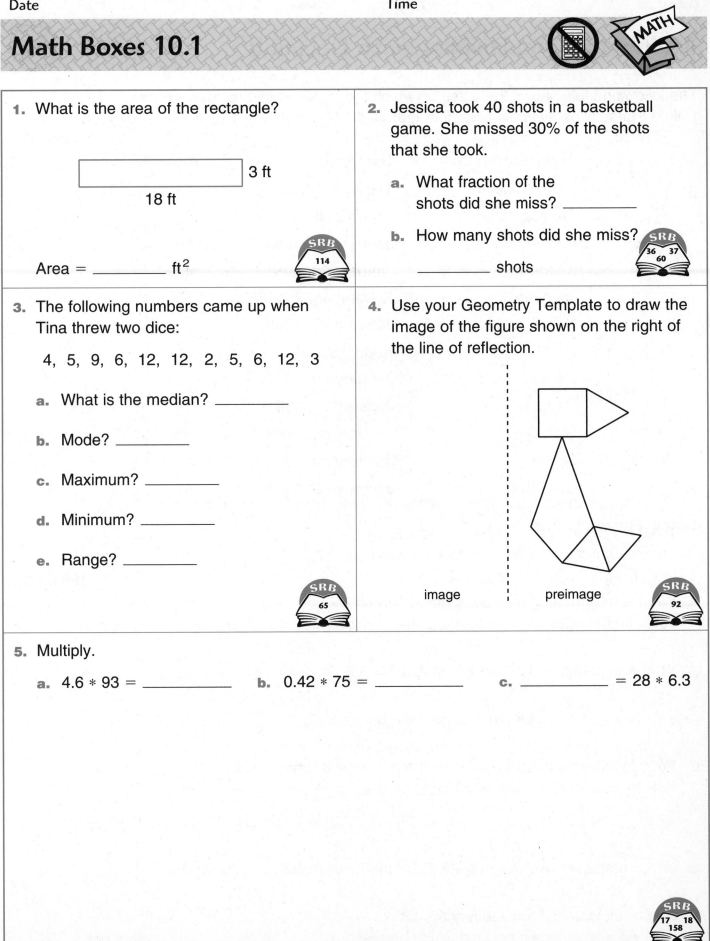

1. What is the area of the rectangle?

   3 ft

   18 ft

   Area = _____ ft^2

   SRB 114

2. Jessica took 40 shots in a basketball game. She missed 30% of the shots that she took.

   a. What fraction of the shots did she miss? _____

   b. How many shots did she miss?

   _____ shots

   SRB 36 37 60

3. The following numbers came up when Tina threw two dice:

   4, 5, 9, 6, 12, 12, 2, 5, 6, 12, 3

   a. What is the median? _____

   b. Mode? _____

   c. Maximum? _____

   d. Minimum? _____

   e. Range? _____

   SRB 65

4. Use your Geometry Template to draw the image of the figure shown on the right of the line of reflection.

   image   preimage

   SRB 92

5. Multiply.

   a. 4.6 * 93 = _____   b. 0.42 * 75 = _____   c. _____ = 28 * 6.3

   SRB 17 18 158

Presidential Information

The following table shows the dates on which the most recent presidents of the United States were sworn in and their ages at the time they were sworn in.

President	Date Sworn In	Age
F.D. Roosevelt	March 4, 1933	51
Truman	April 12, 1945	60
Eisenhower	January 20, 1953	62
Kennedy	January 20, 1961	43
Johnson	November 22, 1963	55
Nixon	January 20, 1969	56
Ford	August 9, 1974	61
Carter	January 20, 1977	52
Reagan	January 20, 1981	69
G.H. Bush	January 20, 1989	64
Clinton	January 20, 1993	46
G.W. Bush	January 20, 2001	54

1. What is the median age (the middle age) of
   the presidents at the time they were sworn in? _____ years

2. What is the range of their ages (the difference
   between the ages of the oldest and the youngest)? _____ years

3. Who was president for the longest time? _____

4. Who was president for the shortest time before 2001? _____

5. Presidents are elected to serve for 1 term. A term lasts 4 years.
   Which presidents served only 1 term or less than 1 term?

   _____

6. Which president was sworn in 28 years after Roosevelt? _____

7. Roosevelt was born on January 30, 1882.
   If he were alive today, how old would he be? _____ years old

Use with Lesson 10.2.

Math Boxes 10.2

1. Write five names for −12.

a. _____

b. _____

c. _____

d. _____

e. _____

SRB 58 131

2. Complete the table with equivalent names.

Fraction	Decimal	Percent
		29%
	0.30	
$\frac{8}{10}$		
		90%

SRB 59 60

3. Write the ordered pair for each point plotted on the coordinate grid.

A (____ , ____)

B (____ , ____)

C (____ , ____)

D (____ , ____)

E (____ , ____)

SRB 124

4. Find the flag of Hungary on page 215 in the World Tour section of your *Student Reference Book.* Be sure to consider color as you answer the following questions:

a. Does this flag have a vertical line of symmetry? _____

b. Does it have a horizontal line of symmetry? _____

SRB 95

5. Divide.

a. 91.6 ÷ 4 = _____

b. _____ = 84.6 ÷ 3

c. _____ = 128.4 ÷ 6

SRB 21 22 158

1. Jillian and Lara estimated the weight of their cat. Circle the most reasonable estimate.

   2 pounds

   10 pounds

   50 pounds

   SRB 154

2. Measure angle *ART*.

   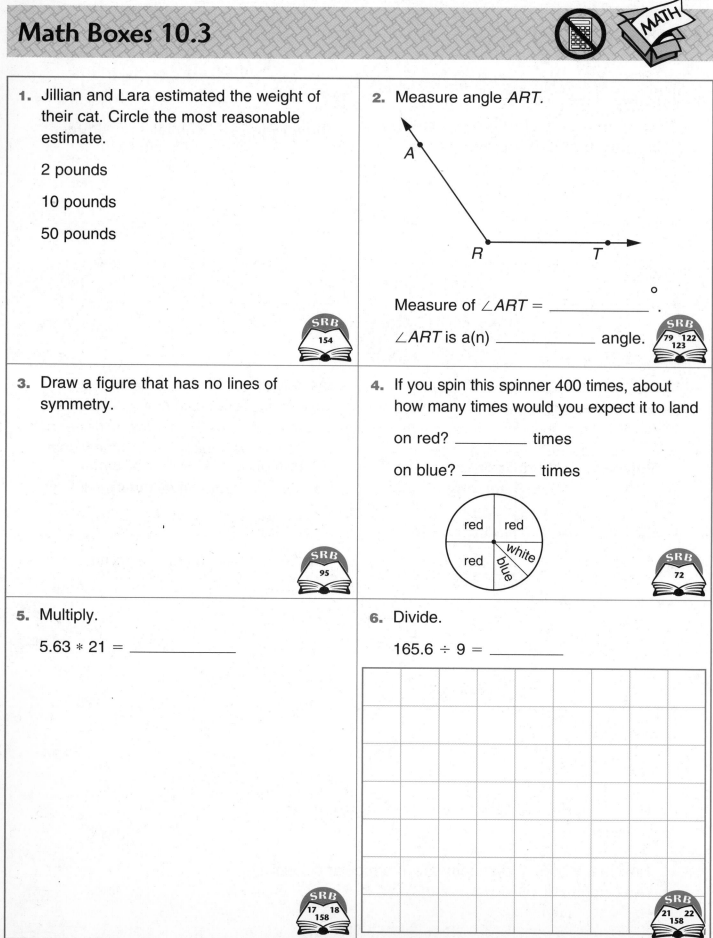

   Measure of ∠*ART* = _____ °.

   ∠*ART* is a(n) _____ angle.

   SRB 79 122 123

3. Draw a figure that has no lines of symmetry.

   SRB 95

4. If you spin this spinner 400 times, about how many times would you expect it to land

   on red? _____ times

   on blue? _____ times

   SRB 72

5. Multiply.

   5.63 * 21 = _____

   SRB 17 18 158

6. Divide.

   165.6 ÷ 9 = _____

   SRB 21 22 158

Use with Lesson 10.3.

Line Symmetry

You will need *Math Masters,* pages 156–159.

1. The drawings on *Math Masters,* page 156 are only half-pictures. Figure out what each whole picture would show. Then use a transparent mirror to complete each picture. Use the recessed side of the mirror to draw the line of reflection.

2. The pictures on *Math Masters,* page 157 are symmetric.

   a. Use the transparent mirror to draw the line of symmetry for the bat and the turtle.

   b. Cut out the other three pictures and find their lines of symmetry by folding.

   c. Which picture has two lines of symmetry? _____

3. Cut out each polygon on *Math Masters,* pages 158 and 159. Find all the lines of symmetry for each polygon. Record the results below.

Polygon	Number of Lines of Symmetry
A	
B	
C	
D	
E	

Polygon	Number of Lines of Symmetry
F	
G	
H	
I	
J	

4. Study the results in the tables above.

   a. How many lines of symmetry are in a regular pentagon (Polygon I)?   _____ lines

   b. How many lines of symmetry are in a regular hexagon (Polygon J)?   _____ lines

   c. How many lines of symmetry are in a regular octagon? (An octagon has 8 sides.)   _____ lines

1. What is the area of the rectangle?

22 m

15 m

Area = _____ m²

2. Tyler missed 20% of the problems on his social studies test. There were 30 problems on the test.

   a. What fraction of the problems did he miss? _____

   b. How many problems did he miss?

   _____ problems

3. Use the following list of numbers to answer the questions.

   7, 8, 24, 8, 9, 17, 17, 8, 12, 13, 19

   a. What is the median? _____

   b. Mode? _____

   c. Maximum? _____

   d. Minimum? _____

   e. Range? _____

4. Use your Geometry Template to draw the image of the figure shown on the top of the line of reflection.

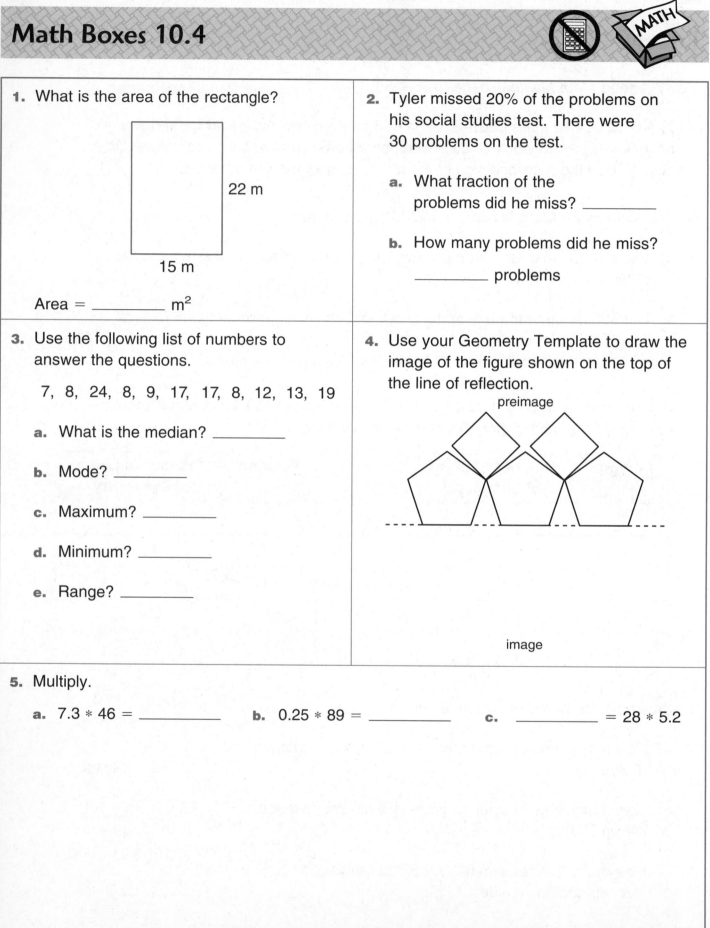

preimage

image

5. Multiply.

   a. $7.3 * 46 =$ _____
   b. $0.25 * 89 =$ _____
   c. _____ $= 28 * 5.2$

Frieze Patterns

1. Extend the following frieze patterns. Use a straightedge and your transparent mirror to help you.

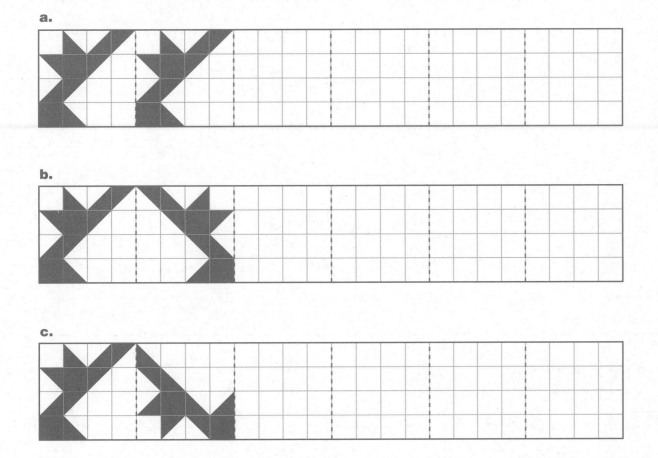

   a.

   b.

   c.

2. Create your own frieze patterns. Make a design in part of the first box. Then repeat the design, using reflections, slides, or rotations. When you have finished, you may want to color or shade your frieze pattern.

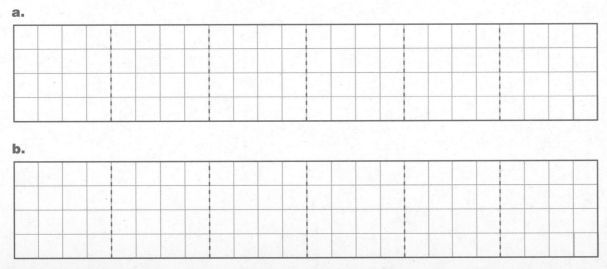

   a.

   b.

Multiplying and Dividing with Decimals

Multiply. Show your work.

1. 9.6 * 36 = _____   **2.** 84 * 0.75 = _____   **3.** _____ = 4.62 * 53

Divide. Show your work.

4. 9)‾38.7‾   **5.** 94.4 ÷ 4 = _____   **6.** _____ = 377.6 / 8

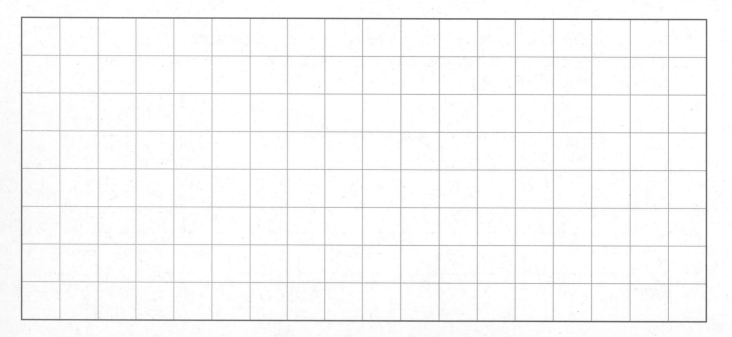

Math Boxes 10.5

---

1. Write five names for −73.

a. _____

b. _____

c. _____

d. _____

e. _____

2. Complete the table with equivalent names.

Fraction	Decimal	Percent
		31%
	0.10	
$\frac{4}{16}$		
		5%

---

3. Write the ordered pair for each point plotted on the coordinate grid.

A (____ , ____)

B (____ , ____)

C (____ , ____)

D (____ , ____)

E (____ , ____)

4. Find two flags on page 215 in the World Tour section of your *Student Reference Book* that have both horizontal and vertical symmetry. (Remember that the colors must also be symmetric.)

a. _____

b. _____

---

5. Divide.

a. 74.8 ÷ 4 = _____

b. _____ = 88.5 ÷ 3

c. _____ = 193.6 ÷ 8

Credits/Debits Game Recording Sheets

Game 1

Recording Sheet

	Start	Change	End/Next Start
1	+$10		
2			
3			
4			
5			
6			
7			
8			
9			
10			

Game 2

Recording Sheet

	Start	Change	End/Next Start
1	+$10		
2			
3			
4			
5			
6			
7			
8			
9			
10			

Number line: -22 -21 -20 -19 -18 -17 -16 -15 -14 -13 -12 -11 -10 -9 -8 -7 -6 -5 -4 -3 -2 -1 0 1 2 3 4 5 6 7 8 9 10 11 12 13 14 15 16 17 18 19 20 21 22

Use with Lesson 10.6.

Math Boxes 10.6

---

1. Hannah and Joshua weighed their mother. Circle the most reasonable weight.

50 pounds

150 pounds

500 pounds

2. Measure angle *RUG*.

Measure of ∠*RUG* = _____ °.

∠*RUG* is a(n) _____ angle.

---

3. Draw a figure that has exactly 1 line of symmetry.

4. If you spin this spinner 540 times, about how many times would you expect it to land

on red? _____ times

on blue? _____ times

---

5. Multiply.

9.46 * 42 = _____

6. Divide.

180.5 ÷ 5 = _____

Time to Reflect

1. Think of the front of your house or apartment building. Does it have a line of symmetry? Draw a picture of it in the space below. Then explain why it is or is not symmetric.

My Home

_____

_____

_____

_____

_____

2. This unit was your first introduction to positive and negative numbers. Did you find them easy or hard to work with? Why?

_____

_____

_____

_____

_____

_____

1. What is the area of the rectangle?

9 cm

4.8 cm

Area = _____ cm²

2. Write five names for −214.

a. _____

b. _____

c. _____

d. _____

e. _____

3. For each animal, circle the most reasonable estimate of its weight.

a. raccoon   > 500 pounds   < 500 pounds   about 500 pounds

b. tiger   > 500 pounds   < 500 pounds   about 500 pounds

c. blue whale   > 500 pounds   < 500 pounds   about 500 pounds

d. giraffe   > 500 pounds   < 500 pounds   about 500 pounds

e. squirrel   > 500 pounds   < 500 pounds   about 500 pounds

4. Draw a rectangle whose area is 12 square centimeters and whose perimeter is 16 centimeters.

Estimating Weights in Grams and Kilograms

A nickel weighs about 5 grams (5 g).
A liter of soda pop weighs about 1 kilogram (1 kg).

In Problems 1–8, circle a possible weight for each object.

1. A dog might weigh about

    20 kg   200 kg   2,000 kg

2. A single-serving can of soup might weigh about

    4 g   40 g   400 g

3. A newborn baby might weigh about

    3 kg   30 kg   300 kg

4. An adult ostrich might weigh about

    1.5 kg   15 kg   150 kg

5. A bowling ball might weigh about

    5 kg   50 kg   500 kg

6. A basketball might weigh about

    0.6 kg   6 kg   60 kg

7. The weight limit in an elevator might be about

    100 kg   1,000 kg   10,000 kg

8. A pencil might weigh about

    2.5 g   25 g   250 g

Date _____ Time _____

Metric and Customary Weight

The number line below has ounces on the top and grams on the bottom.
It shows, for example, that 7 ounces are about equal to 200 grams.

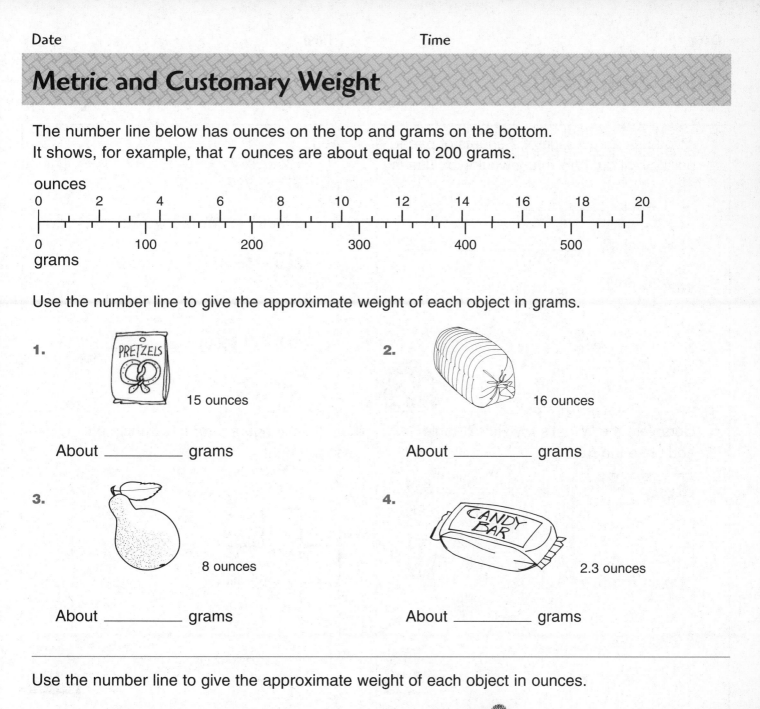

ounces

0   2   4   6   8   10   12   14   16   18   20

0   100   200   300   400   500

grams

Use the number line to give the approximate weight of each object in grams.

1.   15 ounces

About _____ grams

2.   16 ounces

About _____ grams

3.   8 ounces

About _____ grams

4.   2.3 ounces

About _____ grams

Use the number line to give the approximate weight of each object in ounces.

5.   100 grams

About _____ ounces

6.   500 grams

About _____ ounces

7.   140 grams

About _____ ounces

8.   454 grams

About _____ ounces

Use with Lesson 11.1.

Math Boxes 11.1

1. If you use an average of 7 sheets of paper per day, about how many would you use in

a. 1 week? _____

b. 4 weeks? _____

c. 52 weeks? _____

SRB
149 150

2. Add.

a. $-\$75 + \$25 =$ _____

b. $-\$45 + (-\$30) =$ _____

c. $-\$60 + \$60 =$ _____

d. $\$55 + (-\$25) =$ _____

e. $\$300 + (-\$100) =$ _____

3. Complete the "What's My Rule?" table and state the rule.

Rule

in	out
411	298
212	99
	647
555	

SRB
142 143

4. Draw the figure after it is translated to the right.

SRB
92 93

5. Find the solution of each open sentence.

a. $6 * y = 72$ $y =$ _____

b. $9 = 81 / a$ $a =$ _____

c. $98 + s = 425$ $s =$ _____

d. $m - 708 = 292$ $m =$ _____

SRB
128

6. a. What capital city is located at approximately 33°S latitude and 70°W longitude?

_____

b. In which country is the city located?

_____

c. On which continent is the city located?

_____

SRB
216 217

Geometric Solids

3-dimensional geometric shapes like these are also called **geometric solids.**

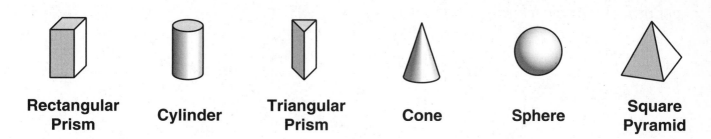

Rectangular Prism **Cylinder** **Triangular Prism** **Cone** **Sphere** **Square Pyramid**

Look around the classroom. Try to find examples of the geometric solids pictured above. Draw a picture of each. Then write its name (for example: book).

Example of rectangular prism:	Example of cylinder:	Example of triangular prism:
Name of object: _____	Name of object: _____	Name of object: _____
Example of cone:	Example of sphere:	Example of square pyramid:
Name of object: _____	Name of object: _____	Name of object: _____

Modeling a Rectangular Prism

After you make a rectangular prism,
answer the questions below.

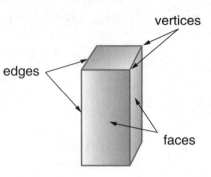

1. How many faces does your rectangular prism have? _____ faces

2. How many of these faces are formed by rectangles? _____ faces

3. How many of these faces are formed by squares? _____ faces

4. Pick one of the faces. How many other faces are parallel to it? _____ face(s)

5. How many edges does your rectangular prism have? _____ edges

6. Pick an edge. How many other edges are parallel to it? _____ edges

7. How many vertices does your rectangular prism have? _____ vertices

8. Write T (true) or F (false) for each of the following statements
about the rectangular prism you made.

   a. _____ It has no curved surfaces.   b. _____ It is a cylinder.

   c. _____ All of the faces are polygons.   d. _____ It has exactly four faces.

   e. _____ All of the edges are parallel.   f. _____ It has more vertices than faces.

Challenge

9. Draw a picture of your rectangular
prism. You can show hidden edges
with dashed lines (- - - - -).

Math Boxes 11.2

1. A cinnamon raisin bagel has about 230 calories. About how many calories are in one dozen bagels?

_____ calories

SRB
149 150

2. Round each number to the nearest tenth.

a. 2.34 _____

b. 0.68 _____

c. 14.35 _____

d. 1.62 _____

e. 5.99 _____

SRB
156 157

3. Draw the figure after it is rotated clockwise a $\frac{1}{4}$ turn.

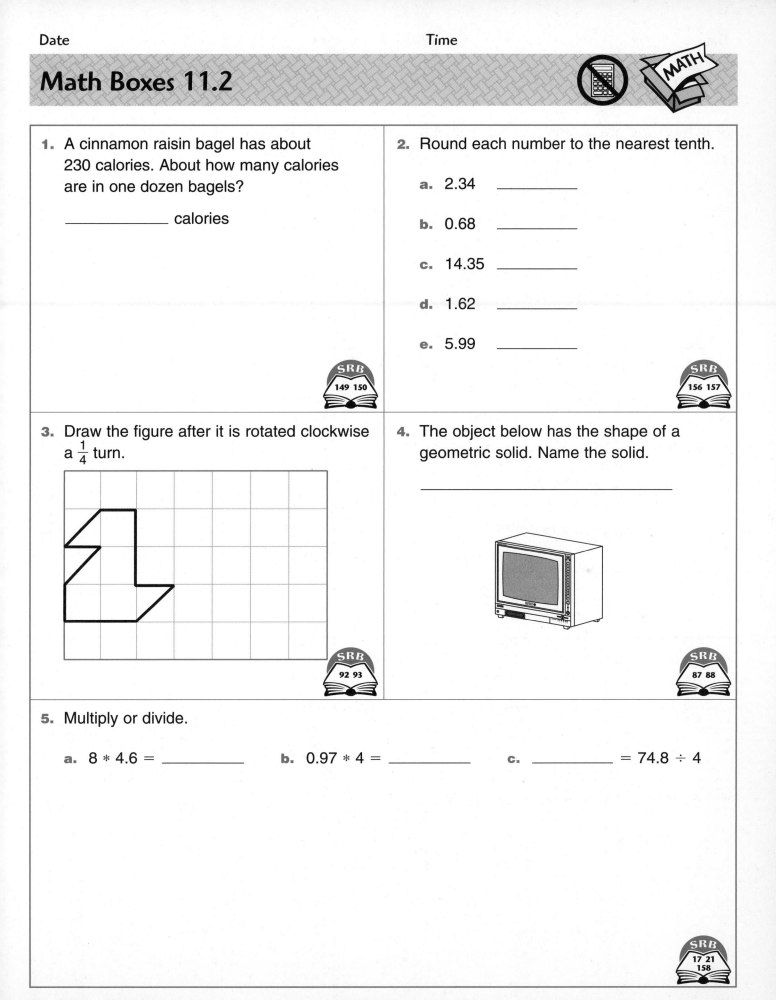

SRB
92 93

4. The object below has the shape of a geometric solid. Name the solid.

_____

SRB
87 88

5. Multiply or divide.

a. 8 * 4.6 = _____   b. 0.97 * 4 = _____   c. _____ = 74.8 ÷ 4

SRB
17 21
158

Construction of Polyhedrons

Polyhedrons are geometric solids with flat surfaces formed by polygons.

For each problem below—

- Decide what the polyhedron should look like.
- Use straws and twist-ties to model the polyhedron.
- Answer the questions about the polyhedron.

Look at page 88 of the *Student Reference Book* if you need help with the name.

1. I am a polyhedron.
   I have 5 faces.
   Four of my faces are formed by triangles.
   One of my faces is a square.

   a. After you make me, draw a picture of me.

   b. What am I?   _____

   c. How many corners (vertices) do I have?   _____

   d. What shape is my base?   _____

2. I am a polyhedron.
   I have 4 faces.
   All of my faces are formed by equilateral triangles.
   All of my faces are the same size.

   a. After you make me, draw a picture of me.

   b. What am I?   _____

   c. How many corners (vertices) do I have?   _____

   d. What shape is my base?   _____

Math Boxes 11.3

1. Gum costs $0.80 per pack. What is the cost of

   a. 4 packs of gum? _____

   b. 10 packs of gum? _____

   c. 16 packs of gum? _____

2. Add.

   a. −$20 + $30 = _____

   b. −$35 + (−$35) = _____

   c. −$15 + (−$40) = _____

   d. $10 + (−$25) = _____

   e. $0 + (−$100) = _____

3. Complete the "What's My Rule?" table and state the rule.

Rule

in	out
313	99
229	15
806	
	887

4. Draw the figure after it is translated to the right.

5. Find the solution of each open sentence.

   a. $x / 8 = 7$ $x =$ _____

   b. $194 = a - 5$ $a =$ _____

   c. $54 = 6 * s$ $s =$ _____

   d. $105 - y = 45$ $y =$ _____

6. a. What capital city is located at approximately 33°N latitude and 7°W longitude?

   _____

   b. In which country is the city located?

   _____

   c. On which continent is the city located?

   _____

The World's Largest Foods

Food	Weight	Date	Location
apple	3 pounds, 11 ounces	not available	Linton, England
broccoli	35 pounds	1993	Palmer, Alaska
bowl of spaghetti	605 pounds	August 16, 1998	London, England
Chinese dumpling	1,058 pounds, 3 ounces	July 5, 1997	Hong Kong
gyro	1.03 tons	July 1998	Zurich, Switzerland
hamburger	2.5 tons	August 5, 1989	Seymour, Wisconsin
ice cream sundae	22.59 tons	July 24, 1988	Alberta, Canada
jelly doughnut	1.5 tons	January 21, 1993	Utica, New York
pineapple	17 pounds, 12 ounces	1994	Ais Village, Papua New Guinea
pumpkin	1,092 pounds	October 3, 1998	Ontario, Canada

Source: Guinness World Records 2000, Millennium Edition

Use the information in the table to solve the following problems.

1. The largest apple weighed _____ ounces.

2. The largest Chinese dumpling weighed _____ ounces.

3. How much more did the largest broccoli weigh than the largest pineapple?

   _____ pounds, _____ ounces

4. A ton is equal to 2,000 pounds. The largest hamburger weighed about

   _____ pounds.

5. Which two foods each weighed about $\frac{1}{2}$ ton?

   _____ and _____

6. A kilogram is a little more than 2 pounds. Which of the foods weighed about

   275 kilograms? _____

Math Boxes 11.4

1. Jake can ride his bike 5 miles in 40 minutes. At this rate, how long does it take him to ride 1 mile?

   _____ minutes

2. Round each number to the nearest tenth.

   a. 3.46 _____

   b. 0.71 _____

   c. 4.35 _____

   d. 9.60 _____

   e. 22.89 _____

3. Draw the figure after it is rotated counterclockwise a $\frac{1}{4}$ turn.

4. The object below has the shape of a geometric solid. Name the solid.

   _____

   CORN

5. Multiply or divide.

   a. $9 * 6.8 =$ _____

   b. $7.56 * 4 =$ _____

   c. _____ $= 63.5 \div 5$

Cube-Stacking Problems

Imagine that each picture at the bottom of this page and on the next page shows a box, partially filled with cubes. The cubes in each box are the same size. Each box has at least one stack of cubes that goes all the way up to the top.

Your task is to find the number of cubes needed to completely fill each box.

Record your answers in the table below.

Table of Volumes						
Placement of Cubes	**Box 1**	**Box 2**	**Box 3**	**Box 4**	**Box 5**	**Box 6**
Number of cubes needed to cover the bottom						
Number of cubes in the tallest stack (Be sure to count the bottom cube.)						
Number of cubes needed to fill the box						

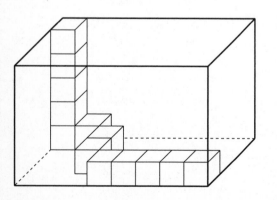

Box 1

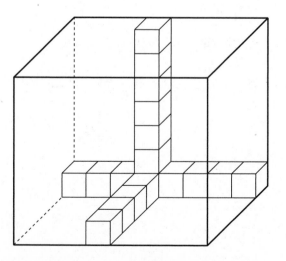

Box 2

Cube-Stacking Problems (cont.)

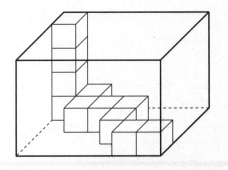

Box 3

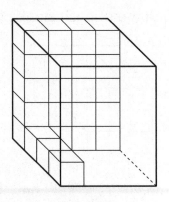

Box 4

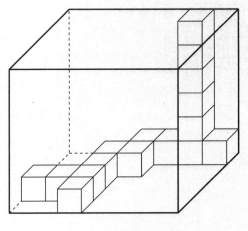

Box 5

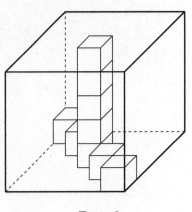

Box 6

Formula for the volume of a rectangular prism:

_____

B is the **area** of a base.

h is the height from that base.

Volume units are cubic units.

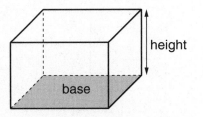

height

base

Cube-Stacking Problems (cont.)

Find the volume of each stack of centimeter cubes.

1.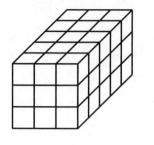

Volume = _____ cm³

2.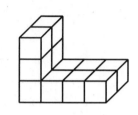

Volume = _____ cm³

3.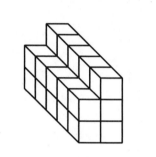

Volume = _____ cm³

4.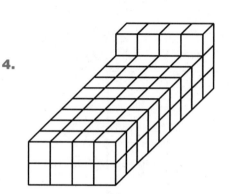

Volume = _____ cm³

5.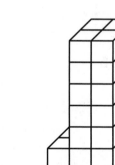

Volume = _____ cm³

6.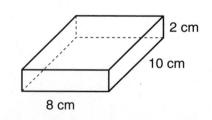

Volume = _____ cm³

7.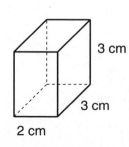

3 cm

3 cm

2 cm

Volume = _____ cm³

8.

2 cm

10 cm

8 cm

Volume = _____ cm³

Math Boxes 11.5

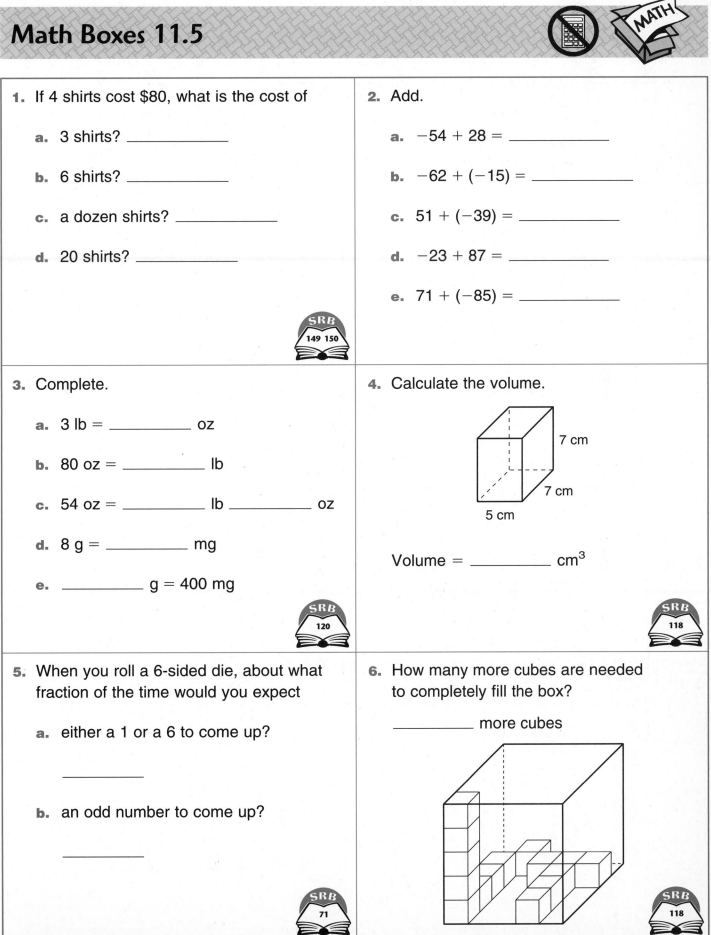

1. If 4 shirts cost $80, what is the cost of

   a. 3 shirts? _____

   b. 6 shirts? _____

   c. a dozen shirts? _____

   d. 20 shirts? _____

SRB
149 150

2. Add.

   a. $-54 + 28 =$ _____

   b. $-62 + (-15) =$ _____

   c. $51 + (-39) =$ _____

   d. $-23 + 87 =$ _____

   e. $71 + (-85) =$ _____

3. Complete.

   a. 3 lb = _____ oz

   b. 80 oz = _____ lb

   c. 54 oz = _____ lb _____ oz

   d. 8 g = _____ mg

   e. _____ g = 400 mg

SRB
120

4. Calculate the volume.

7 cm

7 cm

5 cm

Volume = _____ cm^3

SRB
118

5. When you roll a 6-sided die, about what fraction of the time would you expect

   a. either a 1 or a 6 to come up?

   _____

   b. an odd number to come up?

   _____

SRB
71

6. How many more cubes are needed to completely fill the box?

_____ more cubes

SRB
118

Credits/Debits Game (Advanced Version) Recording Sheets

Game 1

	Start	Change		End, and next start
		Addition or Subtraction	Credit or Debit	
1	+ $10			
2				
3				
4				
5				
6				
7				
8				
9				
10				

Game 2

	Start	Change		End, and next start
		Addition or Subtraction	Credit or Debit	
1	+ $10			
2				
3				
4				
5				
6				
7				
8				
9				
10				

Use with Lesson 11.6.

Math Boxes 11.6

1. Two cups of flour are needed to make about 20 medium-size peanut butter cookies. How many cups of flour will you need to make about

   a. 40 cookies? _____ cups

   b. 60 cookies? _____ cups

   c. 50 cookies? _____ cups

2. Round each number to the nearest tenth.

   a. 8.99 _____

   b. 0.06 _____

   c. 21.76 _____

   d. 1.53 _____

   e. 0.92 _____

3. Draw the figure after it is rotated clockwise a $\frac{1}{2}$ turn.

4. The object below has the shape of a geometric solid. Name the solid.

   _____

5. Multiply or divide.

   a. $6 * 32.9 =$ _____

   b. $98.7 \div 3 =$ _____

   c. _____ $= 55.2 \div 12$

Converting Measurements

Math Message

1 pint = _____ cups

1 quart = _____ pints

1 half-gallon = _____ quarts

1 gallon = _____ quarts

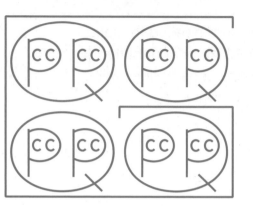

How can the picture above help you remember how many cups are in a pint, how many pints are in a quart, and how many quarts are in a gallon?

Rice Consumption

1. Round your answer to the nearest ounce.

   One cup of dry (uncooked) rice weighs about _____ ounces.

2. Use the answer in Problem 1 to complete the following:

   a. 1 pint of rice weighs about _____ ounces.

   b. 1 quart of rice weighs about _____ ounces.

   c. 1 gallon of rice weighs about _____ ounces.

   d. 1 gallon of rice weighs about _____ pounds. (1 pound = 16 ounces)

3. On average, a family of 4 in Japan eats about 40 pounds of rice a month.

   a. That's about how many **pounds** a year? _____ pounds

   b. How many **gallons**? _____ gallons

4. On average, a family of 4 in the United States eats about 88 pounds of rice a year.
   That's about how many gallons a year? _____ gallons

5. On average, a family of 4 in Thailand eats about 3 gallons of rice a week.

   a. That's about how many gallons a year? _____ gallons

   b. How many pounds? _____ pounds

Use with Lesson 11.7.

Math Boxes 11.7

1. If you travel at an average speed of 50 miles per hour, how far will you travel in

   a. 3 hours? _____

   b. $\frac{1}{2}$ hour? _____

   c. $2\frac{1}{2}$ hours? _____

   d. 12 hours? _____

2. Add.

   a. $-46 + 20 =$ _____

   b. $-23 + (-18) =$ _____

   c. $33 + (-17) =$ _____

   d. $-45 + 66 =$ _____

   e. $27 + (-40) =$ _____

3. Complete.

   a. 7 lb = _____ oz

   b. 32 oz = _____ lb

   c. 72 oz = _____ lb _____ oz

   d. 12 g = _____ mg

   e. _____ g = 600 mg

4. Calculate the volume.

   2 cm

   5 cm

   9 cm

   Volume = _____ cm³

5. When you roll a 6-sided die, about what fraction of the time would you expect

   a. a number less than 5 to come up?

   _____

   b. an even number to come up?

   _____

6. How many more cubes are needed to completely fill the box?

   _____ more cubes

Time to Reflect

1. In your own words, describe the difference between a 2-dimensional figure and a 3-dimensional shape.

   _____

   _____

   _____

   _____

   _____

2. You played the *Credits/Debits Game* again in this unit. You also played the *Credits/Debits Game* (Advanced Version). What do you think is the most difficult part of these games? What is the easiest part? Explain your answers.

   _____

   _____

   _____

   _____

   _____

3. What were some of your least favorite activities in this unit? Explain why you disliked them.

   _____

   _____

   _____

   _____

   _____

4. What were some of your favorite activities in this unit? Explain why you liked them.

   _____

   _____

   _____

   _____

   _____

Use with Lesson 11.8.

Math Boxes 11.8

1. If you use the telephone an average of 4 times per day, about how many times would you use it in

   a. 1 week? _____ times

   b. 4 weeks? _____ times

   c. 52 weeks? _____ times

2. A cup of orange juice has about 110 calories. About how many calories are in a quart of orange juice?

   _____ calories

3. Candy bars cost $0.55 each. What is the cost of

   a. 4 candy bars? _____

   b. 10 candy bars? _____

   c. 18 candy bars? _____

4. If you walk at an average speed of 3.5 miles per hour, how far will you travel in

   a. 2 hours? _____ miles

   b. 6 hours? _____ miles

   c. $\frac{1}{2}$ hour? _____ miles

5. Michelle can run 5 miles in 35 minutes. At this rate, how long does it take her to run 1 mile?

   _____ minutes

6. Round each number to the nearest tenth.

   a. 5.87 _____

   b. 0.32 _____

   c. 9.65 _____

   d. 3.40 _____

   e. 93.29 _____

Rates

1. While at rest, a typical student in my class blinks _____ times in one minute.

2. While reading, a typical student in my class blinks _____ times in one minute.

3. List as many examples of rates as you can.

   _____

   _____

   _____

   _____

   _____

   _____

   _____

   _____

   _____

4. Find at least two examples of rates in your *Student Reference Book.* (*Hint:* Look at pages 214 and 243.)

   _____

   _____

   _____

   _____

   _____

   _____

   _____

   _____

   _____

Use with Lesson 12.1.

Math Boxes 12.1

1. I am a whole number. Use the clues to figure out what number I am.

Clue 1: If you multiply me by 2, I become a number greater than 20 and less than 40.

Clue 2: If you multiply me by 6, I end in 8.

Clue 3: I am an odd number.

What number am I? _____

SRB 8

2. Calculate the volume.

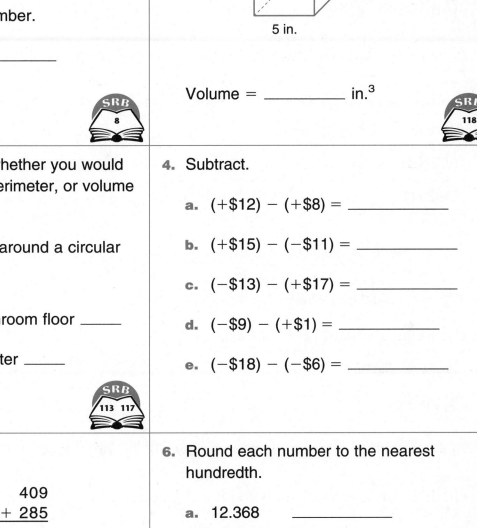

6 in.

8 in.

5 in.

Volume = _____ in.^{3}

SRB 118

3. Write A, P, or V to tell whether you would need to find the area, perimeter, or volume in each situation.

a. finding the distance around a circular track _____

b. buying tile for a bathroom floor _____

c. filling a pool with water _____

SRB 113 117

4. Subtract.

a. $(+\$12) - (+\$8) =$ _____

b. $(+\$15) - (-\$11) =$ _____

c. $(-\$13) - (+\$17) =$ _____

d. $(-\$9) - (+\$1) =$ _____

e. $(-\$18) - (-\$6) =$ _____

5. Solve.

a.
```
  109
-  48
```

b.
```
  409
+ 285
```

SRB 9 11

6. Round each number to the nearest hundredth.

a. 12.368 _____

b. 234.989 _____

c. 1.225 _____

d. 12.304 _____

e. 0.550 _____

SRB 156 157

Rate Tables

For each problem, fill in the rate table. Then answer the question below the table.

1. Bill's new car can travel about 35 miles on 1 gallon of gasoline.

Gasoline mileage: 35 miles per gallon

Miles	35							
Gallons	1	2	3	4	5	6	7	8

At this rate, about how far can the car travel on 7 gallons of gas? _____ miles

2. Jennifer received an allowance of $8 in 4 weeks.

Allowance: $8 in 4 weeks

Dollars				8				
Weeks	1	2	3	4	5	6	7	8

At this rate, how much allowance did Jennifer receive per week? $_____

3. A gray whale's heart beats 24 times in 3 minutes.

Gray whale's heart rate: 24 beats in 3 minutes

Heartbeats			24					
Minutes	1	2	3	4	5	6	7	8

At this rate, how many times does a gray whale's heart beat in 2 minutes? _____ times

4. Mr. Johnson paid $1.80 for 3 pounds of grapes.

Cost of grapes: 3 pounds for $1.80

Pounds	1	2	3	4	5	6	7	8
Dollars			1.80					

At this rate, how much do 5 pounds of grapes cost? $_____

Use with Lesson 12.2.

Math Boxes 12.2

1. Write 5 names for 8.01.

a. _____

b. _____

c. _____

d. _____

e. _____

SRB
131

2. Find the solution of each open sentence.

a. $m + 40 = -60$ $m =$ _____

b. $55 + q = 40$ $q =$ _____

c. $(-23) + s = 0$ $s =$ _____

d. $p + (-36) = -80$ $p =$ _____

SRB
128

3. Complete.

a. 3 gal = _____ qt

b. 4 qt = _____ pt

c. 6 pt = _____ c

d. _____ qt = _____ pt

e. _____ gal = 16 qt

SRB
117

4. **a.** Complete the table.

Number of Pizzas	1	2		4	5
Number of Servings	3		9	12	

b. How many pizzas are needed for 21 servings?

_____ pizzas

SRB
142

5. Divide.

$319.2 \div 6 =$ _____

SRB
21 22
158

6. Calculate.

a. 10% of 460 = _____

b. 5% of 120 = _____

c. 40% of _____ = 4

d. _____% of 20 = 6

e. _____% of 92 = 46

SRB
36 37

Do These Numbers Make Sense?

Math Message

It is estimated that the average lifetime of a person living in the United States is about 75 years.

About how many days are there in an average lifetime?   About _____ days

About how many hours is that?   About _____ hours

Use the data from the Math Message to help you answer the following questions:

1. It is estimated that in an average lifetime of
   75 years, a person sleeps for about 214,000 hours.
   At that rate, about how many hours *per day* does
   a person sleep?   _____ hours per day

   Does this number make sense to you?   _____

2. It is estimated that in an average lifetime, a person
   watches TV for about 105,000 hours. At that rate, about
   how many hours *per day* does a person watch TV? _____ hours per day

   Does this number make sense to you?   _____

3. It is estimated that in an average lifetime, a person
   laughs about 540,000 times. At that rate, about how
   many times *per day* does a person laugh?   _____ times per day

   Does this number make sense to you?   _____

4. It is estimated that in an average lifetime, a person takes about
   95,000,000 breaths. Does this number make sense to you? Explain.

   _____

   _____

   _____

   _____

Source: The Compass in Your Nose and Other Astonishing Facts about Humans

Use with Lesson 12.3.

Math Boxes 12.3

---

1. I am a whole number. Use the clues to figure out what number I am.

    Clue 1: I am less than 100.

    Clue 2: The sum of my digits is 4.

    Clue 3: Half of me is an odd number.

    What number am I? _____

2. Calculate the volume.

    9 m

    5 m

    1 m

    Volume = _____ m^3

---

3. Write A, P, or V to tell whether you would need to find the area, perimeter, or volume in each situation.

    a. buying tile for a bedroom ceiling _____

    b. buying a wedding ring _____

    c. buying dirt for a potted plant _____

4. Subtract.

    a. $(+\$8) - (+\$3) =$ _____

    b. $(+\$7) - (-\$2) =$ _____

    c. $(-\$8) - (+\$11) =$ _____

    d. $(-\$5) - (+\$3) =$ _____

    e. $(-\$12) - (-\$5) =$ _____

---

5. Solve.

    a.   1,000
          − 397

    b.   678
          + 930

6. Round each number to the nearest hundredth.

    a. 4.568   _____

    b. 0.123   _____

    c. 6.155   _____

    d. 9.780   _____

    e. 15.349   _____

Product Testing

Some magazines written for young people ask their readers to test many different kinds of products. The results of the tests are then published in the magazines to help readers make wise buying decisions. For example, in one issue of *Zillions,* the child's version of *Consumer Reports* magazine, 44 of the magazine's readers taste-tested several brands of potato chips. The readers considered taste, cost, and nutritional value as they tried to decide which brand was the "best buy." In another issue of *Zillions,* a team of testers compared 37 brands of peanut butter in their search for the best product.

When a reader wrote to the magazine to complain about a board game she had bought, the magazine sent board games to young people in every part of the country. Testers were asked to play each game several times and then to report on what they liked and disliked about the game.

1. If you were testing a board game, what are some of the features you would look for?

   _____

   _____

   _____

   _____

2. When readers of the magazine tested potato chips, they considered taste, cost, and nutritional value in determining the best chip. Which of these factors is the most important to you? Why?

   _____

   _____

   _____

   _____

3. What is a **consumer**? Be prepared to share your definition with the class.

   _____

   _____

   _____

   _____

Unit Prices

Solve the unit price problems below. Complete the tables if it is helpful to do so.

1. A 12-ounce can of soda pop costs 60 cents. The unit price is _____ per ounce.

Dollars				0.60
Ounces	1	3	6	12

2. A 4-pound bunch of bananas costs $1.16. The unit price is _____ per pound.

Dollars				1.16
Pounds	1	2	3	4

3. A 5-pound bag of apples costs $1.90. The unit price is _____ per pound.

Dollars					1.90
Pounds	1	2	3	4	5

4. Three pounds of salmon cost $21.00.

a. The unit price is _____ per pound.

b. What is the cost of 7 pounds of salmon? _____

c. What is the cost of $9\frac{1}{2}$ pounds of salmon? _____

Dollars			21.00			
Pounds	1	2	3	4	7	$9\frac{1}{2}$

5. *Snickeroo* candy bars come in packages of 25 and cost $3.50 per package. *Yummy* candy bars come in packages of 30 and cost $3.60 per package. Which is the better buy? _____ candy bars

Explain. _____

_____

_____

_____

Rates

1. A mole can dig a tunnel 300 feet long in one night.
   How far could a mole dig in one week?   About _____ feet

2. An elephant may eat 500 pounds of hay and drink
   60 gallons of water in one day.

   a. About how many pounds of hay could an
      elephant eat per week?   About _____ pounds

   b. About how many gallons of water could an
      elephant drink per week?   About _____ gallons

3. The bottle-nosed whale can dive to a depth
   of 3,000 feet in 2 minutes. About how many
   feet is that per second?   About _____ feet per second

4. A good milking cow will give up to 1,500 gallons of milk in a year.

   a. About how many gallons is that in 3 months?   About _____ gallons

   b. About how many *quarts* is that in 3 months?   About _____ quarts

Challenge

5. Sloths spend up to 80 percent of their lives sleeping. Not only is a sloth extremely
   sleepy, it is also very slow. A sloth travels on the ground at a speed of about
   7 feet per minute. In the trees, its speed is about 15 feet per minute.

   a. After one hour, how much farther would
      a sloth have traveled in the trees than on
      the ground (if it didn't stop to sleep)?   About _____ feet

   b. About how long would it take a sloth to
      travel 1 mile on the ground? (*Hint:* There
      are 5,280 feet in a mile.)   About _____ minutes,

                                                      or _____ hours

334

Math Boxes 12.4

1. Write 5 names for 3.16.

    a. _____

    b. _____

    c. _____

    d. _____

    e. _____

2. Find the solution of each open sentence.

    a. $t + 30 = -120$ $t =$ _____

    b. $75 + n = 20$ $n =$ _____

    c. $16 + b = 0$ $b =$ _____

    d. $c + (-61) = -97$ $c =$ _____

3. Complete.

    a. 7 gal = _____ qt

    b. 3 qt = _____ pt

    c. 8 pt = _____ c

    d. _____ qt = 32 pt

    e. _____ gal = 40 qt

4. a. Complete the table.

Number of Cookies		72			540
Number of Packages	1	2	9	12	

    b. How many cookies are in 8 packages?

    _____ cookies

5. Divide.

    $325.2 \div 4 =$ _____

6. Calculate.

    a. 10% of 520 = _____

    b. 5% of 180 = _____

    c. 60% of _____ = 12

    d. _____ % of 30 = 15

    e. _____ % of 35 = 14

Unit Pricing

Math Message

1. Use your calculator to divide. Write down what the calculator displays for each quotient. Your teacher will tell you how to fill in the answer spaces for "cents."

   a. $9.52 ÷ 7 = $____._____, or _____ cents

   b. $1.38 ÷ 6 = $____._____, or _____ cents

   c. $0.92 ÷ 8 = $____._____, or _____ cents

   d. $0.98 ÷ 6 = $____._____, or about _____ cents

   e. $1.61 ÷ 9 = $____._____, or about _____ cents

2. A package of 6 candy bars costs $2.89. What is the price of 1 candy bar?   _____ cents

3. A 15-ounce bottle of shampoo costs $3.89. What is the price per ounce?   _____ cents

4. Brand A: a box of 16 crayons for 80 cents
   Brand B: a box of 32 crayons of the same kind for $1.28

   Which box is the better buy?   _____

   Why? _____

   _____

   _____

   _____

Challenge

5. A store sells a 3-pound can of coffee for $7.98 and a 2-pound can of the same brand for $5.98. You can use a coupon worth 70 cents toward the purchase of the 2-pound can. If you use the coupon, which is the better buy, the 3-pound can or the 2-pound can? Explain your answer.

   _____

   _____

   _____

   _____

More Unit-Pricing Problems

1. A 15-ounce box of cereal costs $3.60.
   What is the price per ounce?   _____

2. 20 pounds of potatoes cost $7.40.
   What is the price per pound?   _____

3. One pound of sliced turkey costs $5.44.
   What is the price per ounce? (*Hint:* 1 pound = 16 ounces)   _____

4. A store sells a 4-pack of AA batteries for $2.40. It sells a
   6-pack of the same kind for $3.30. Which box is the better buy?

   _____

   Why? _____

   _____

   _____

   _____

5. A 6-ounce bag of potato chips costs $1.50. A 14-ounce bag costs the same amount
   per ounce as the 6-ounce bag.

   a. How much does the 14-ounce bag cost? $_____

   Explain your answer. _____

   _____

   _____

   b. Which is the better buy—the 6-ounce bag or the 14-ounce bag?

   _____

   _____

   _____

1. Multiply. Show your work.

46 * 231 = _____

SRB 17 18

2. a. Complete the table.

Number of Cups		32			272
Number of Gallons	1	2	9	12	

b. How many cups are in 5 gallons?

_____ cups

SRB 117

3. An average 10-year-old drinks about 20 *gallons* of soft drinks a year.

a. At that rate, about how many *cups* does a 10-year-old drink in a month?

_____ cups

b. Does this number make sense to you?

_____

_____

_____

SRB 117

4. Subtract. Do not use a calculator.

a. (+$9) − (+$4) = _____

b. (+$8) − (−$3) = _____

c. (−$7) − (+$15) = _____

d. (−$6) − (+$1) = _____

e. (−$14) − (−$9) = _____

5. Calculate the volume.

8 ft

8 ft

8 ft

Volume = _____ ft³

SRB 118

6. A 3-ounce bag of corn chips costs $0.65. A 14-ounce bag of corn chips costs $2.79.

a. What is the price per ounce of each bag? (Round to the nearest cent.)

3-oz bag: _____

14-oz bag: _____

b. Which bag of chips is the better buy?

_____

SRB 149 156

Use with Lesson 12.5.

Looking Back on the World Tour

Math Message

It is time to complete the World Tour.

1. Fly to Washington, D.C., and then travel to your hometown. Mark the final leg of the tour on the Route Map on *Math Journal 2,* pages 346 and 347.

2. What is the total distance you have traveled? _____ miles

3. The airline has given you a coupon for every 5,000 miles you have traveled. Suppose you did all your traveling by plane on the same airline. How many coupons have you earned on the World Tour? _____ coupons

4. You can trade in 5 coupons for one free round-trip ticket to fly anywhere in the continental United States. How many round-trip tickets have you earned on the World Tour? _____ round-trip tickets

Refer to "My Country Notes" in your journals (*Math Journal 1,* pages 182–187 and *Math Journal 2,* pages 348–355) as you answer the following questions.

5. If you could travel all over the world for a whole year, what information would you need in order to plan your trip?

_____

_____

_____

_____

_____

_____

_____

_____

Looking Back on the World Tour (cont.)

6. To which country would you most like to travel in your lifetime? Explain your answer.

_____

_____

_____

_____

_____

7. On your travels, you would have the opportunity to learn about many different cultures. What would you want to share with people from other countries about _your_ culture?

_____

_____

_____

_____

_____

8. What are some things about the World Tour that you did not enjoy?

_____

_____

_____

_____

9. What are some things you have enjoyed on the World Tour?

_____

_____

_____

_____

Rates

1. According to a 1990 survey, men under 25 years old spend an average of 53 minutes a day arranging their hair and clothes. At this rate, about how much time do they spend arranging hair and clothes in a week?

   About _____

2. People drink an average of $2\frac{1}{2}$ quarts of water a day. At this rate, about how many quarts of water do they drink in 2 weeks?

   About _____

3. On average, Americans eat about 19 pounds of pasta per year.

   a. At this rate, about how many pounds of pasta would they eat in 23 years?

      About _____

   b. Does this number seem reasonable to you?

      _____

4. Thirty-six buses would be needed to carry the passengers and crew of three 747 jumbo jets.

   a. Fill in the rate table.

Buses			36	180	252
Jets	1	2	3		

   b. How many jets would be needed to carry the passengers on 264 buses?

      _____

5. A man in India grew one of his thumbnails until it was 114 centimeters long. Fingernails grow about 2.5 centimeters each year. At this rate, about how many years did it take the man in India to grow his thumbnail?

   About _____

6. At Martha's Vineyard in Massachusetts, the waves are wearing away the cliffs along the coast at a rate of about 5 feet per year. About how long would it take for 10 yards to be worn away?

   About _____

1. Write 5 names for 2.75.

   a. _____

   b. _____

   c. _____

   d. _____

   e. _____

2. Find the solution of each open sentence.

   a. $y + (-8) = -23$ $y =$ _____

   b. $12 + j = -5$ $j =$ _____

   c. $35 + r = 25$ $r =$ _____

   d. $c + (-115) = -144$ $c =$ _____

3. Give 3 other names for each measure.

   a. 1 gal

   _____ _____ _____

   b. 1 qt

   _____ _____ _____

   c. 1 pt

   _____ _____ _____

4. a. Complete the table. Ignore leap years.

Number of Days	365			
Number of Years	1	2	9	12

   b. How many days are in 7 years?

   _____ days

5. Divide.

   $384.5 \div 5 =$ _____

6. Calculate.

   a. 10% of 860 = _____

   b. 5% of 220 = _____

   c. 75% of _____ = 12

   d. _____% of 87 = 43.5

   e. _____% of 60 = 18

Use with Lesson 12.6.

Time to Reflect

1. Answer one of the following questions about your role in the eye-blinking experiment.

   a. If you were in the group that collected the eye-blinking data, which part of your job did you find the most difficult? Which part was the easiest?

   _____

   _____

   _____

   b. If you were in the group that was watched, what did you think your partner was doing? Did you guess what kind of data he or she was collecting?

   _____

   _____

   _____

2. Do you find that a rate table helps you solve rate problems? Why or why not?

   _____

   _____

   _____

   _____

3. What do you think is required in order to be considered a "smart" consumer?

   _____

   _____

   _____

   _____

4. Give five examples of rates.

   _____

   _____

   _____

   _____

Math Boxes 12.7

1. Multiply. Show your work.

79 * 405 = _____

2. a. Complete the table.

Number of Inches				144	192	324
Number of Feet	1	2	9	12		

b. How many inches are in 11 feet?

_____ inches

3. It is estimated that in an average lifetime of 75 years, a person takes about 50,000 trips in a car.

a. At that rate, about how many times a day does a person ride in a car?

_____ times

b. Does this number make sense to you?

_____

4. Subtract. Do not use a calculator.

a. $(-\$75) - (+\$25) =$ _____

b. $(-\$45) - (-\$30) =$ _____

c. $(-\$60) - (+\$60) =$ _____

d. $\$55 - (-\$25) =$ _____

e. $\$300 - (-\$100) =$ _____

5. Calculate the volume.

Volume = _____ cm³

6. A 10-ounce can of peas costs $0.55. A 16-ounce can of peas costs $1.19.

a. What is the price per ounce for each can? (Round to the nearest cent.)

10-oz can: _____

16-oz can: _____

b. Which can of peas is the better buy?

_____

Use with Lesson 12.7.

My Route Log

Date	Country	Capital	Air distance from last capital	Total distance traveled so far
	1 U.S.A.	Washington, D.C.	███	
	2 Egypt	Cairo		
	3			
	4			
	5			
	6			
	7			
	8			
	9			
	10			
	11			
	12			
	13			
	14			
	15			
	16			
	17			
	18			
	19			
	20			

Route Map

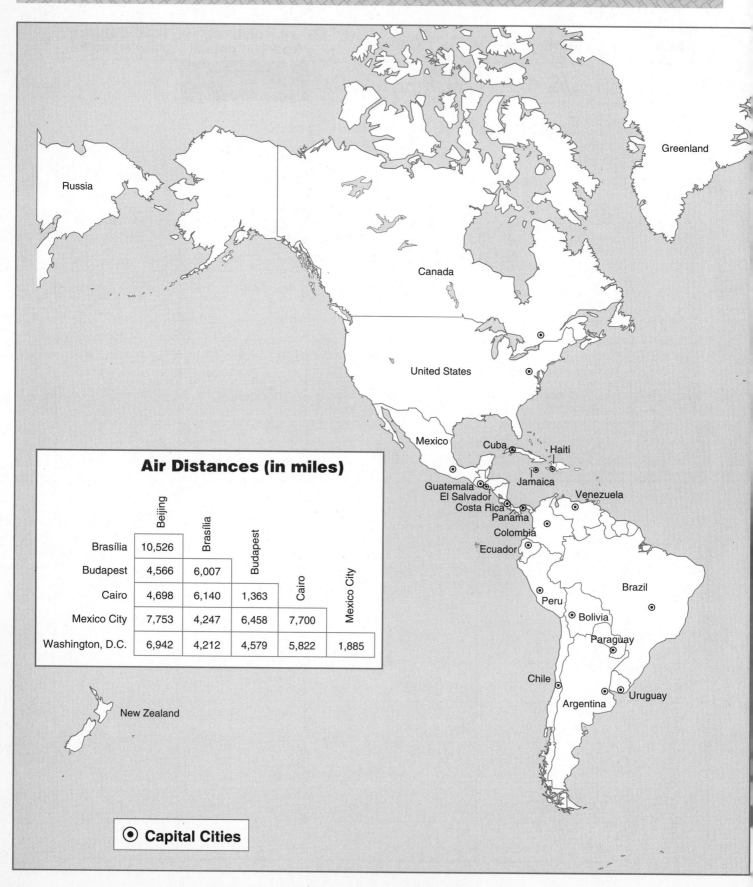

Air Distances (in miles)

	Beijing	Brasília	Budapest	Cairo	Mexico City
Brasília	10,526				
Budapest	4,566	6,007			
Cairo	4,698	6,140	1,363		
Mexico City	7,753	4,247	6,458	7,700	
Washington, D.C.	6,942	4,212	4,579	5,822	1,885

⊙ **Capital Cities**

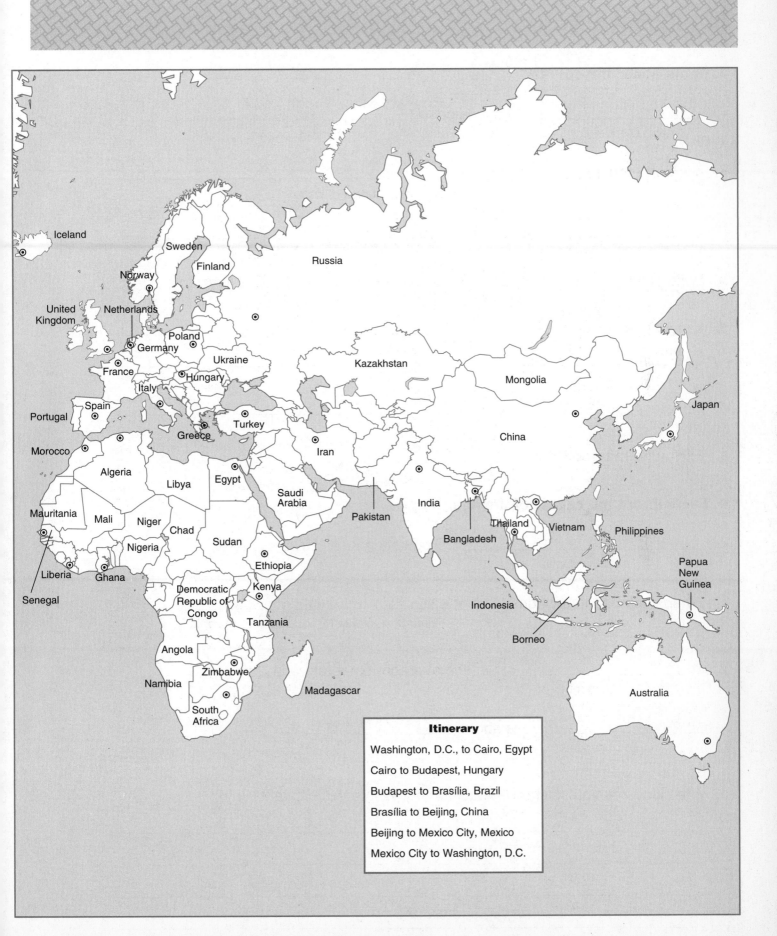

Iceland
Sweden
Finland
Russia
Norway
United Kingdom
Netherlands
Poland
Germany
Ukraine
France
Hungary
Italy
Spain
Greece
Turkey
Portugal
Morocco
Algeria
Libya
Egypt
Iran
Kazakhstan
Mongolia
Japan
China
Saudi Arabia
Mauritania
Mali
Niger
Chad
Pakistan
India
Thailand
Vietnam
Philippines
Nigeria
Sudan
Bangladesh
Liberia
Ghana
Senegal
Democratic Republic of Congo
Ethiopia
Kenya
Tanzania
Indonesia
Borneo
Papua New Guinea
Angola
Zimbabwe
Namibia
Madagascar
South Africa
Australia

Itinerary

Washington, D.C., to Cairo, Egypt

Cairo to Budapest, Hungary

Budapest to Brasília, Brazil

Brasília to Beijing, China

Beijing to Mexico City, Mexico

Mexico City to Washington, D.C.

My Country Notes

A. Facts about the country

_____ is located in _____ .
  name of country   name of continent

1. It is bordered by _____
                                              countries, bodies of water

   _____ .

2. Population: _____   Area: _____ square miles

3. Languages spoken: _____

   _____

4. Monetary unit: _____

5. Exchange rate (optional): 1 _____ = _____

B. Facts about the capital of the country

_____   Population: _____
  name of capital

1. When it is noon in my hometown, it is _____ in _____ .
                                          time (A.M. or P.M.?)   name of capital

2. In _____ , the average temperature in
        month

   _____ is about _____ °F.
     name of capital

3. What kinds of clothes should I pack for my visit to this capital? Why?

   _____

   _____

   _____

   _____

My Country Notes (cont.)

4. Turn to the Route Map found on journal pages 346 and 347.
   Draw a line from the last city you visited to the capital of this country.

5. If your class is using the Route Log, take journal page 345 or *Math Masters,* page 38 and record the information.

6. Can you find any facts on pages 246–249 in your *Student Reference Book* that apply to this country? For example, is one of the 10 tallest mountains in the world located in this country? List all the facts you can find.

_____

_____

_____

_____

_____

c. **My impressions about the country**

Do you know anyone who has visited or lived in this country? If so, ask that person for an interview. Read about the country's customs and about interesting places to visit there. Use encyclopedias, travel books, the travel section of a newspaper, or library books. Try to get brochures from a travel agent. Then describe below some interesting things you have learned about this country.

_____

_____

_____

_____

_____

_____

_____

_____

_____

My Country Notes

A. Facts about the country

_____ is located in _____.
name of country   name of continent

1. It is bordered by _____
                                              countries, bodies of water

   _____.

2. Population: _____   Area: _____ square miles

3. Languages spoken: _____

   _____

4. Monetary unit: _____

5. Exchange rate (optional): 1 _____ = _____

B. Facts about the capital of the country

_____   Population: _____
        name of capital

1. When it is noon in my hometown, it is _____ in _____.
                                                    time (A.M. or P.M.?)   name of capital

2. In _____, the average temperature in
              month

   _____ is about _____ °F.
        name of capital

3. What kinds of clothes should I pack for my visit to this capital? Why?

   _____

   _____

   _____

   _____

My Country Notes (cont.)

4. Turn to the Route Map found on journal pages 346 and 347.
   Draw a line from the last city you visited to the capital of this country.

5. If your class is using the Route Log, take journal page 349 or *Math Masters,* page 38 and record the information.

6. Can you find any facts on pages 246–249 in your *Student Reference Book* that apply to this country? For example, is one of the 10 tallest mountains in the world located in this country? List all the facts you can find.

_____

_____

_____

_____

_____

c. My impressions about the country

Do you know anyone who has visited or lived in this country? If so, ask that person for an interview. Read about the country's customs and about interesting places to visit there. Use encyclopedias, travel books, the travel section of a newspaper, or library books. Try to get brochures from a travel agent. Then describe below some interesting things you have learned about this country.

_____

_____

_____

_____

_____

_____

_____

_____

My Country Notes

A. Facts about the country

_____ is located in _____.
    name of country   name of continent

1. It is bordered by _____
                                countries, bodies of water

_____.

2. Population: _____   Area: _____ square miles

3. Languages spoken: _____

_____

4. Monetary unit: _____

5. Exchange rate (optional): 1 _____ = _____

B. Facts about the capital of the country

_____   Population: _____
    name of capital

1. When it is noon in my hometown, it is _____ in _____.
                                     time (A.M. or P.M.?)   name of capital

2. In _____, the average temperature in
             month

_____ is about _____ °F.
   name of capital

3. What kinds of clothes should I pack for my visit to this capital? Why?

_____

_____

_____

_____

                  Use with Lesson 10.3.

My Country Notes (cont.)

4. Turn to the Route Map found on journal pages 346 and 347.
   Draw a line from the last city you visited to the capital of this country.

5. If your class is using the Route Log, take journal page 345 or *Math Masters,* page 38 and record the information.

6. Can you find any facts on pages 246–249 in your *Student Reference Book* that apply to this country? For example, is one of the 10 tallest mountains in the world located in this country? List all the facts you can find.

   _____

   _____

   _____

   _____

   _____

c. **My impressions about the country**

   Do you know anyone who has visited or lived in this country? If so, ask that person for an interview. Read about the country's customs and about interesting places to visit there. Use encyclopedias, travel books, the travel section of a newspaper, or library books. Try to get brochures from a travel agent. Then describe below some interesting things you have learned about this country.

   _____

   _____

   _____

   _____

   _____

   _____

   _____

   _____

   _____

My Country Notes

A. Facts about the country

_____ is located in _____.
      name of country   name of continent

1. It is bordered by _____
                                 countries, bodies of water

_____.

2. Population: _____   Area: _____ square miles

3. Languages spoken: _____

_____

4. Monetary unit: _____

5. Exchange rate (optional): 1 _____ = _____

B. Facts about the capital of the country

_____   Population: _____
      name of capital

1. When it is noon in my hometown, it is _____ in _____.
                                     time (A.M. or P.M.?)   name of capital

2. In _____, the average temperature in
         month

_____ is about _____ °F.
    name of capital

3. What kinds of clothes should I pack for my visit to this capital? Why?

_____

_____

_____

_____

                                                          Use with Lesson 11.1.

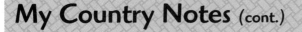

My Country Notes (cont.)

4. Turn to the Route Map found on journal pages 346 and 347.
   Draw a line from the last city you visited to the capital of this country.

5. If your class is using the Route Log, take journal page 345 or *Math Masters,* page
   38 and record the information.

6. Can you find any facts on pages 246–249 in your *Student Reference Book* that
   apply to this country? For example, is one of the 10 tallest mountains in the world
   located in this country? List all the facts you can find.

   _____

   _____

   _____

   _____

   _____

c. My impressions about the country

Do you know anyone who has visited or lived in this country? If so, ask
that person for an interview. Read about the country's customs and about
interesting places to visit there. Use encyclopedias, travel books, the travel section
of a newspaper, or library books. Try to get brochures from a travel agent. Then
describe below some interesting things you have learned about this country.

   _____

   _____

   _____

   _____

   _____

   _____

   _____

   _____

   _____

Equivalent Names for Fractions

Fraction	Equivalent Fractions	Decimal	Percent
$\frac{0}{2}$		0	0%
$\frac{1}{2}$	$\frac{2}{4}, \frac{3}{6}$		
$\frac{2}{2}$		1	100%
$\frac{1}{3}$			
$\frac{2}{3}$			
$\frac{1}{4}$			
$\frac{3}{4}$			
$\frac{1}{5}$			
$\frac{2}{5}$			
$\frac{3}{5}$			
$\frac{4}{5}$			
$\frac{1}{6}$			
$\frac{5}{6}$			
$\frac{1}{8}$			
$\frac{3}{8}$			
$\frac{5}{8}$			
$\frac{7}{8}$			

Use with Lesson 7.6.

Equivalent Names for Fractions (cont.)

Fraction	Equivalent Fractions	Decimal	Percent
$\frac{1}{9}$			
$\frac{2}{9}$			
$\frac{4}{9}$			
$\frac{5}{9}$			
$\frac{7}{9}$			
$\frac{8}{9}$			
$\frac{1}{10}$			
$\frac{3}{10}$			
$\frac{7}{10}$			
$\frac{9}{10}$			
$\frac{1}{12}$			
$\frac{5}{12}$			
$\frac{7}{12}$			
$\frac{11}{12}$			

Fraction Cards 1

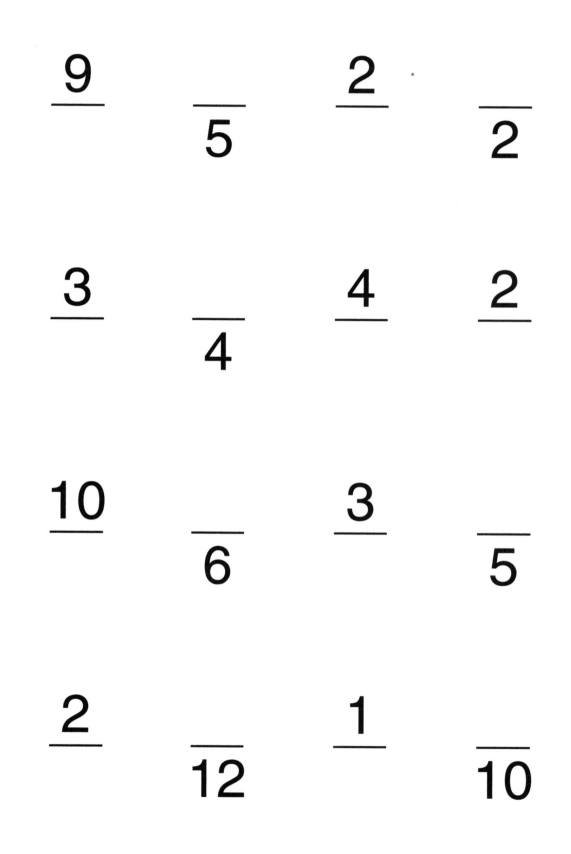

Date Time

Fraction Cards 2

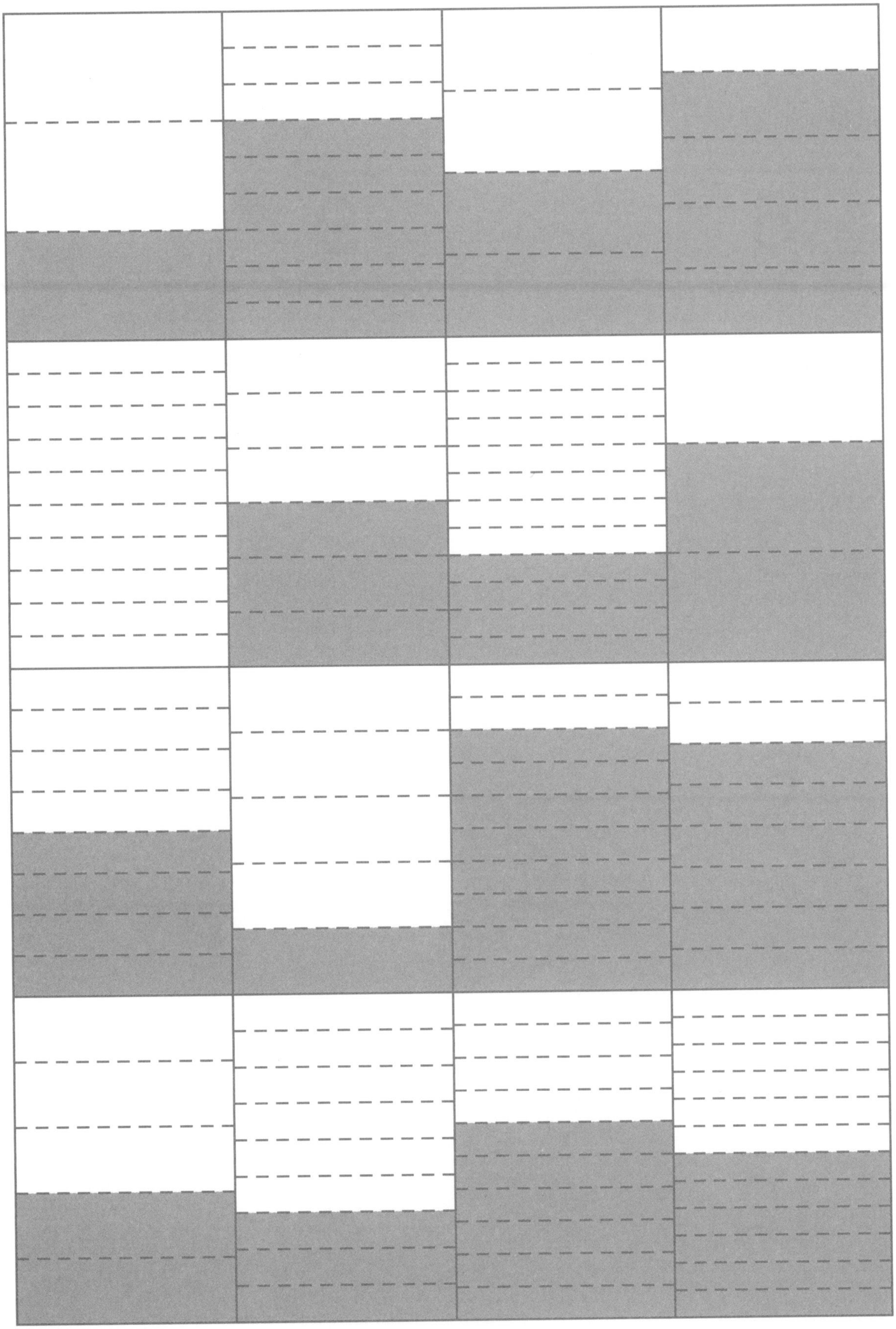

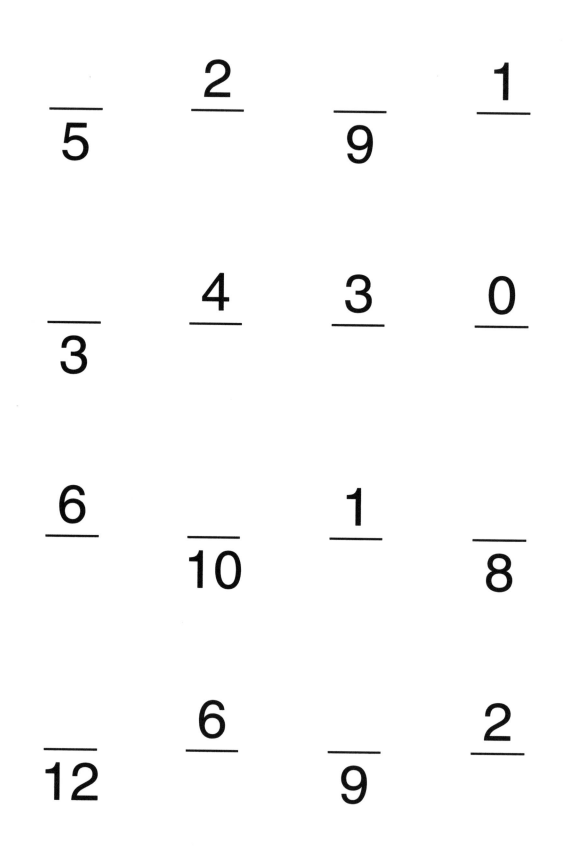

Fraction/Percent Tiles

10%	20%	25%	30%
40%	50%	60%	70%
75%	80%	90%	100%
$\frac{1}{2}$	$\frac{1}{4}$	$\frac{3}{4}$	$\frac{1}{5}$
$\frac{2}{5}$	$\frac{3}{5}$	$\frac{4}{5}$	$\frac{1}{10}$
$\frac{3}{10}$	$\frac{7}{10}$	$\frac{9}{10}$	$\frac{2}{2}$

P	P	P	P
P	P	P	P
P	P	P	P
F	F	F	F
F	F	F	F
F	F	F	F

Decimal Tiles

0.10	0.20	0.25	0.30
0.40	0.50	0.60	0.70
0.75	0.80	0.90	1

D D D D

D D D D

D D D D